FISH Handbook for Biological Wastewater Treatment

FISH Handbook for Biological Wastewater Treatment

Identification and quantification of microorganisms in activated sludge and biofilms by FISH

Edited by
Per Halkjær Nielsen, Holger Daims and Hilde Lemmer

IWA Publishing
London • New York

Published by **IWA Publishing**
Alliance House
12 Caxton Street
London SW1H 0QS, UK
Telephone: +44 (0)20 7654 5500
Fax: +44 (0)20 7654 5555
Email: publications@iwap.co.uk
Web: www.iwapublishing.com

First published 2009

Cover images provided by Simon Jon McIlroy, Jeppe Lund Nielsen, and Kilian Stoecker

Cover design by www.designforpublishing.co.uk

Typeset in India by Alden Prepress Services Private Limited.

Index provided by Alden Prepress Services Private Limited.

Printed by Page Bros Ltd, Norwich, UK.

British Library Cataloguing in Publication Data
A CIP catalogue record for this book is available from the British Library

Library of Congress Cataloging-in-Publication Data
A catalog record for this book is available from the Library of Congress

ISBN 13: 9781843392316

Table of contents

List of contributors

Editors:

Professor Per Halkjær Nielsen
Department of Biotechnology, Chemistry and Environmental Engineering,
Aalborg University, Sohngaardsholmsvej 49,
DK-9000 Aalborg, Denmark
Fax: 45 9814 1808
Phone: 45 9940 8503
Email: phn@bio.aau.dk

Dr. Hilde Lemmer
Bavarian Environment Agency, Kaulbachstr. 37,
D-80539 Munich, Germany
Phone: +49 89 2180 2783
Fax: +49 89 2800838
Email: lemmer@oec.net

Dr. Holger Daims
Department of Microbial Ecology
Vienna Ecology Centre
University of Vienna, Althanstrasse 14
A-1090 Vienna, Austria
Phone: +43 1 4277 54392
Fax: +43 1 4277 54389
Email: daims@microbial-ecology.net

Other contributors:

Dr. Aviaja A. Hansen
Department of Biotechnology, Chemistry and Environmental Engineering,
Aalborg University, Sohngaardsholmsvej 49,
DK-9000 Aalborg, Denmark
Email: aah@bio.aau.dk

Dr. Caroline Kragelund
Department of Biotechnology, Chemistry and Environmental Engineering,
Aalborg University, Sohngaardsholmsvej 49,
DK-9000 Aalborg, Denmark
Email: ckr@bio.aau.dk

Dr. Frank Maixner
Department of Microbial Ecology, Vienna Ecology Centre
University of Vienna, Althanstrasse 14,
A-1090 Vienna, Austria
Email: maixner@microbial-ecology.net

Simon Jon McIlroy
Biotechnology Research Centre, La Trobe University,
Bendigo, Vic 3552, Australia
Email: simon.j.mcilroy@gmail.com

Artur Tomasz Mielczarek
Department of Biotechnology, Chemistry and Environmental Engineering,
Aalborg University, Sohngaardsholmsvej 49,
DK-9000 Aalborg, Denmark
Email: atm@bio.aau.dk

Dr. Elisabeth Müller
Institute of Water Quality Control, Technical University Munich,
Am Coulombwall, D-85748 Garching, Germany
Email: Lisa.L.Mueller@bv.tum.de

Hien Thi Thu Nguyen
Department of Biotechnology, Chemistry and Environmental Engineering,
Aalborg University, Sohngaardsholmsvej 49,
DK-9000 Aalborg, Denmark
Email: hn@bio.aau.dk

Dr. Jeppe Lund Nielsen
Department of Biotechnology, Chemistry and Environmental Engineering,
Aalborg University, Sohngaardsholmsvej 49,
DK-9000 Aalborg, Denmark
Email: jln@bio.aau.dk

Dr. Margit Schade
Bavarian Environment Agency,
Kaulbachstr. 37, D-80539 Munich, Germany
Email: margit.schade@lfu.bayern.de

Dr. Markus C. Schmid
Department of Microbial Ecology, Vienna Ecology Centre
University of Vienna, Althanstrasse 14,
A-1090 Vienna, Austria
Email: schmid@microbial-ecology.net

Professor Robert J. Seviour
Biotechnology Research Centre, La Trobe University,
Bendigo, Vic 3552, Australia
Email: R.Seviour@latrobe.edu.au

Abbreviations

AOA	Ammonia oxidizing archaea
AOB	Ammonia oxidizing bacteria
BNR	Biological nutrient removal
CLSM	Confocal laser scanning microscope
EBPR	Enhanced biological phosphorus removal
FISH	Fluorescence in situ hybridization
FOV	Microscope fields of view
GALOs	*Gordonia amarae*-like organisms
GAOs	Glycogen-accumulating organisms
NOB	Nitrite oxidizing bacteria
PAOs	Polyphosphate-accumulating organisms
PCR	Polymerase chain reaction
Pi	Inorganic phosphorus
PHA	Polyhydroxyalkanoates
PTLOs	Pine tree-like organisms
rRNA	Ribosomal RNA
TFOs	Tetrad-forming microorganisms
WWTPs	Wastewater treatment plants

1

Introduction

Per Halkjær Nielsen, Holger Daims, and Hilde Lemmer

1.1 IDENTIFICATION OF MICROORGANISMS IN ACTIVATED SLUDGE AND BIOFILMS

Until very recently, culture-dependent methods such as plate count or Most-Probable Number (MPN) counting have widely been used for enumeration and detection of bacteria being relevant to biological wastewater treatment performance. In fact, such standard methods are in many cases still used for effluent quality control, particularly with respect to pathogens and various indicator organisms (e.g. APHA Standard Methods). However, today we know that these methods suffer from severe limitations as from all types of microbes in environmental samples (also pathogens), only a very small fraction is cultivable on media generally applied. Therefore, this approach is prone to lead to serious misinformation. Thus, we strongly advocate for a change to using culture-independent molecular methods in all sorts of microbiological investigations in wastewater treatment plants (WWTPs).

Among the cultivation-independent methods for detection, fluorescence *in situ* hybridization (FISH) with ribosomal RNA (rRNA)-targeted probes (gene probes) is a very powerful tool for identification of microorganisms in activated sludge and biofilm biocenoses from WWTPs. This method is described in detail in Chapters 7 and 8 of this book. Most known functional key microorganisms in wastewater systems can be reliably identified and quantified by this method.

Furthermore, other molecular methods exist, primarily PCR-based methods. Quantitative PCR (q-PCR) is now getting more commonly applied in environmental samples, but the method has several drawbacks compared to FISH. This is due to biases concerning nucleic acid extraction, the PCR reaction, and also the fact that PCR-based approaches do not quantify microbial cells, but measure copy numbers of marker

 FISH Handbook for Biological Wastewater Treatment: Identification and quantification of microorganisms in activated sludge and biofilms by FISH. Edited by Per Halkjær Nielsen, Holger Daims and Hilde Lemmer. ISBN: 9781843392316. Published by IWA Publishing, London, UK.

genes. DNA microarrays carrying rRNA-targeted probes, so-called "phylochips", have a great potential as high-throughput tools for the qualitative detection of hundreds or even thousands of different uncultured microbes in only one experiment. When combined with autoradiography, phylochips become "isotope arrays" useful to track functional traits of microbes such as nitrifying bacteria in WWTP (Adamczyk *et al.*, 2003). However, to date no quantitative phylochip-based assay exists that would allow for observing shifts in the abundances of probe-target microbial populations. This limitation is due to technical problems, such as saturation effects during hybridization, with the microarrays which are very difficult to overcome.

In contrast to the other methods, by using FISH it is possible to observe the morphology and to quantify numbers of bacteria or the equivalent biovolume. Thus, in our opinion, FISH is for the time being the method of choice for detection and quantification of microorganisms in WWTP as detailed in Chapters 7 and 8.

A special case is the identification of filamentous bacteria. These have primarily been identified based on their morphology and simple staining techniques using light microscopy since their first comprehensive description by Eikelboom (Eikelboom, 1975). Several manuals have since been published (e.g. Eikelboom 2000; Jenkins *et al.*, 2004), all of them being based on his original work. However, today it is clear that although some filamentous bacteria can be fairly reliably identified in this way, the majority can not. As described in Chapter 5, we strongly recommend to also apply FISH for the identification of filamentous microorganisms, after having accomplished a preliminary morphological identification using the manuals.

1.2 THE MICROBIOLOGY OF BIOLOGICAL WASTEWATER TREATMENT

Biological treatment of municipal and industrial wastewater worldwide is primarily carried out by the activated sludge (AS) process. New technologies are being developed such as biofilm reactors, membrane bioreactors, sequencing batch reactors, etc., but they basically all derive from the AS process. The common purpose of all these technologies is the use of microorganisms to remove carbon (C), nitrogen (N), phosphorus (P), micropollutants and pathogens.

New interesting more sustainable solutions are appearing. They include for example recovery of nutrients (e.g. P) from wastewater, or conversion of organic waste components to usable, valuable compounds such as bioplastics (polyhydroxyalkanoates, PHA). Conversion of organic waste to energy by methane production during anaerobic digestion has been utilized for decades and these processes are further being developed together with other energy yielding processes such as microbial fuel cells.

Management of these complex microbial systems (or 'microbial resource management' for the new sustainable solutions, Verstraete *et al.*, 2007) relies on a fundamental knowledge about the microbial populations being involved and about the factors that regulate their activity. A reliable identification of the microorganisms involved is fundamental and with the today's toolbox of various culture-independent methods is possible with a high sensitivity and precision. Not only the identity, but also knowledge about their ecophysiology, ecology, and population dynamics is essential. The present methodological approaches range from single cell microbiology (e.g. microautoradiography and FISH-Raman microspectroscopy; Huang *et al.*, 2007), expression of specific functional genes, systems biology (genomics, transcriptomics, and proteomics) to lab-scale reactors and full-scale studies of chemical transformations. In this way we are gaining a rapidly increasing understanding of key microorganisms being involved in many processes and how to affect their presence and activity. However, there is still much to learn about full-scale systems since most studies so far have been carried out in lab-scale and pilot-scale reactors.

Several functional groups of bacteria being involved in the most common treatment processes are now fairly well identified and described. It is primarily bacteria involved in nitrification and to some extent those involved in denitrification, many bacteria involved in the enhanced biological P-removal (EBPR), and most bacteria causing settling problems (bulking) or foam/scum formation. In each functional group, for example the nitrifiers, a limited number of phylogenetic lineages (<10) is encountered in nitrifying plants in general with only a few dominant populations (3–5) being present in a particular plant within the majority of full-scale plants. We should try to avoid the term "species", because a concise species definition is lacking in microbial ecology. Lineages, strains or ecotypes might be equally important for WWTP functioning as "species". The exact microbial community composition in a particular plant depends on wastewater composition, process design, and plant operation, see below. However, in common for most functional groups, the controlling factors determining the community composition is still poorly understood.

Does it matter which bacteria are present in each functional group in a certain treatment plant? This question can in some cases be answered with a clear 'yes', in others 'perhaps' or 'we don't know'. For the filamentous bacteria it is a clear 'yes'. Certain ecotypes cause severe settling properties, others are (in low number) important for sludge flocs as a backbone and thus for the floc structure. A proper identification is essential for the selection of efficient control measures towards the unwanted filamentous bacteria causing settling problems like bulking or foam.

It is more uncertain how important the knowledge about the exact community composition is for the nitrification performance in a certain activated sludge treatment plant, for example. Based on *in situ* observations it has been suggested that the presence of several lineages of ammonia and nitrite oxidizers ensures a more robust and stable system (e.g. Daims *et al.*, 2001c) compared to the presence of only a single lineage from each functional group. But it is less documented what the exact community composition of a certain functional group means for plant stability and operation. Future studies, all based on a reliable identification of the microbial populations, for example by FISH, will show. A related question is whether strain-level microbial diversity influences process stability in WWT. For instance, very closely related nitrite oxidizers, which are slightly different on the genome level, may co-exist in the same plant. Due to a high sequence similarity (or even identity) of their 16S rRNA, the diversity of these strains would easily be overlooked by current approaches using methods such as FISH. However, the genomic differences may result in a biologically significant variety of phenotypes, which respond differently to events such as changes in wastewater composition or bacteriophage attack. With the latest molecular methods such as environmental genomics and deep DNA sequencing at hand, future research will elucidate the importance of such microbial microdiversity from an applied perspective.

Only a few comprehensive studies have been published describing the total community composition from full-scale wastewater treatment plants. They deal with activated sludge plants, such as industrial plants for C-removal or N-removal (Juretschko *et al.*, 2002) or a plant with biological N- and P-removal from a mixture of domestic and industrial wastewater (Kong *et al.*, 2007). However, several studies of specific populations have been carried out in various full-scale plants, referring solely to nitrifiers, denitrifiers or filamentous bacteria, for example. These studies are briefly described in the specific chapters. A summary of the most commonly observed species and genera encountered in WWTPs is shown in Table 1.1.

Table 1.1. Commonly reported microorganisms in wastewater treatment systems.

Functional group/chapter	Commonly reported populations
Nitrifiers (Chapter 2)	
Ammonium oxidizers (AOB)	Genus *Nitrosomonas* (*N. europaea*, *N. eutropha*, *N. mobilis*, and *N. oligotropha*) (class *Betaproteobacteria*) Genus *Nitrosospira* (class *Betaproteobacteria*)
Nitrite oxidizers (NOB)	Genus *Nitrospira* (sublineage 1 and 2) (phylum *Nitrospirae*) Genus *Nitrobacter* (class *Alphaproteobacteria*)
Anammox bacteria	Lineages *Brocadia, Kuenenia*, *Scalindua*, and *Anammoxoglobus* (phylum *Planctomycetes*)
Denitrifiers (Chapter 3)	Genus *Candidatus* Accumulibacter (class *Betaproteobacteria*) Genus *Azoarcus* (class *Betaproteobacteria*) Genus *Curvibacter* (class *Betaproteobacteria*) Genus *Thauera* (class *Betaproteobacteria*) Genus *Zoogloea* (class *Betaproteobacteria*)
Polyphosphate-accumulating organisms (PAOs) (Chapter 4)	Genus *Candidatus* Accumulibacter (class *Betaproteobacteria*) Genus *Tetrasphaera* (phylum *Actinobacteria*)
Glygogen-accumulating organisms (GAOs) (Chapter 4)	Genus *Candidatus* Competibacter (class *Gammaproteobacteria*) Genus *Defluviicoccus* (class *Alphaproteobacteria*)
Filamentous bacteria (Chapter 5)	Species in class *Alphaproteobacteria* Genus *Sphaerotilus* (class *Betaproteobacteria*) Genus *Thiothrix* (*Thiothrix* spp. and type 021N) (class *Gammaproteobacteria*) *Candidatus* Microthrix parvicella (phylum *Actinobacteria*) Genus *Skermania* (phylum *Actinobacteria*) Genus *Gordonia* (phylum *Actinobacteria*) Genus *Rhodococcus* (phylum *Actinobacteria*) Genus *Dietzia* (phylum *Actinobacteria*) Species in phylum and class *Chloroflexi* Genus *Haliscomenobacter* (phylum *Bacteroidetes*) Species in candidate phylum TM7
Others (Chapter 6)	Genus *Candidatus* Epiflobacter spp. (phylum *Bacteroidetes*)

1.3 FACTORS OF IMPORTANCE FOR THE GROWTH OF MICROORGANISMS

The species composition in treatment plants depends on wastewater composition, process design, and plant operation. It is obvious that certain functional groups are dominant only if specific processes are included in the plant's process design (e.g. N-removal or EBPR). As mentioned above, the controlling factors determining the species composition are still poorly understood for many species, so during studies of the microbiology in WWTPs it is important to observe and register the potential factors that may be decisive for the presence of the different species. An overview of such factors is given in Table 1.2 and more advice on these is given elsewhere (e.g. Wilderer *et al.*, 2002).

Table 1.2. Overview of important factors determining the microbial population structure in WWTPs.

Process design	C-removal, C-removal and nitrification
	C- and N-removal (nitrification and denitrification)
	C- and N-removal and EBPR
	Chemical P-precipitation
	Sludge age (total and aerobic)
	Sludge loading
	Temperature level and seasonal variations
	Others
Plant operation	Oxygen concentration
	Mean cell residence time in different tanks
	Addition of chemicals (e.g. Fe/Al salts, polymers)
	Addition of external C (e.g. methanol)
	Biomass content (e.g. suspended solids per liter)
	Others
Treatment plant type	Activated sludge (continuous flow, SBR, ...)
	Biofilter (type, support media, operation, ...)
	Membrane bioreactor
	Others
Wastewater composition	Industrial/domestic
	Soluble/particulate fractions (C, N, P)
	Specific organic compounds (e.g. acetate)
	Micronutrients
	Toxic substances (e.g. metals, sulfides)
	Salinity
	Alkalinity
	pH value
	Others

Most important for activated sludge plants is whether solely C-removal is included in the plant design – meaning that only aerobic tanks are present (besides clarifier), or whether denitrification/EBPR are also included, meaning that anoxic/anaerobic tanks are present in addition to the aerobic ones. The selective pressure due to anoxic/anaerobic tanks substantially changes the population structure. Sludge age (mean cell residence time), which is determined by the sludge loading, is also extremely important. A low sludge age (<5–10 days) may not allow nitrification to occur due to the low growth rates of the nitrifiers, whereas a high sludge age (>20–30 days) is important to obtain full N- and P-removal in temperate climates. Treatment plants running at very high temperatures (>40°C), such as those treating special industrial wastewater, often select for unusual microbial communities.

The composition of the incoming wastewater is another decisive factor for bacterial growth. Industrial wastewater is often less complex than municipal wastewater, meaning fewer microorganisms may dominate in the treatment plants. Furthermore, the soluble fraction is often higher. Industrial wastewater may also not include important nutrients, such as P or N or other micronutrients. Domestic wastewater is usually more complex with a high fraction of particulates and a more balanced ratio of organics and nutrients. Other important factors are temperature, salinity, presence of toxic substances, and pH value. Likewise, the incoming microorganisms may affect the population structure in the treatment plant.

The operation of the plant may also affect the population structure. It is closely interrelated to the process design of the plant. The exact operation of anoxic/anaerobic mean cell residence time, oxygen concentration in aerobic tanks, addition of chemicals, carbon sources, and many other factors may affect the population composition.

The importance of the technology platform being applied to carry out a specific process, such as nitrification, is also poorly understood. Do we get the same nitrifiers in a full-scale plant based on activated sludge with different process configurations (e.g. sequencing batch reactors, continuously stirred reactors or plug-flow reactors), biofilm reactors (e.g. upflow or downflow biofilters, airlift reactors, granules reactors), or membrane bioreactors treating the same wastewater? Few studies have investigated this in detail, but the general impression is that we are often dealing with the same species/groups, although perhaps with slightly different strains/ecotypes (see also above). More studies based on a reliable identification of the microbial populations are needed.

1.4 THE USE OF THIS FISH HANDBOOK

This handbook contains a detailed description of the FISH protocol for identification and quantification of various bacteria typically encountered in biological wastewater treatment. The bacteria included cover several functional groups: nitrifiers, denitrifiers, polyphosphate-accumulating organisms (PAOs), glycogen-accumulating organisms (GAOs), filamentous bacteria involved in bulking or foaming, and some others. They can be found in aerobic or aerobic/anoxic/anaerobic treatment systems based on activated sludge or biofilms applied in various technologies. We have not included bacteria present in digesters (anaerobic digestion), fuel cells or bacteria carrying out more rarely encountered treatment processes such as treatment of S-containing waste (for S^o production) or removal of specific pollutants (e.g. from polluted sites). Some of these groups may be included in future editions of this book.

The handbook does not cover protozoa as the molecular methods are still not ready for a proper detection of these.

Detailed information about the ecophysiology, ecology or management of the different bacteria is outside the scope of this handbook. Such information can be found in the specific literature, manuals, and books (see the different chapters) or in the new IWA book on "Microbial Ecology of Activated Sludge" (Eds. Seviour and Nielsen, 2009).

For each functional group an overview of the identities of the bacteria and their abundances is presented. Extensive tables describe the gene probes to be applied for the detection of the microbes, and phylogenetic trees show the coverage of the various probes. Furthermore, we have included a description of the typical morphologies targeted by the probes and many of these are documented by FISH images in the color figure section (Chapter 9). Based on our own experience we recommend the most suitable probes. Most probes are selected on the following criteria: they were designed based on published full-length sequences, the probes are published, and details can be viewed in probeBase (www.microbial-ecology.net/probebase). Other probes are briefly mentioned in the text. They might be relevant in special cases.

New or improved gene probes for detection of relevant microbes are continually being developed, so we plan ongoing revisions of this book, the first in 1–2 years. Information about new editions and other relevant updates can be found on the web page of the IWA specialist groups (www.iwahq.org). This handbook has been developed in conjunction with the IWA specialist group on 'Activated Sludge Population Dynamics'.

Errors, suggestions for inclusion of other probes, important experiences or other comments are more than welcome as they can be beneficially included in future editions. Please e-mail Per Halkjær Nielsen (phn@bio.aau.dk).

2

The nitrifying microbes: Ammonia oxidizers, nitrite oxidizers, and anaerobic ammonium oxidizers

Holger Daims, Frank Maixner, and Markus C. Schmid

2.1 INTRODUCTION

Nitrification is a key process of nitrogen removal in most municipal and industrial wastewater treatment plants (WWTPs). The term "nitrification" refers to the aerobic, sequential oxidation of ammonia to nitrite and of nitrite to nitrate. These two steps are catalyzed by specialized chemolithoautotrophic prokaryotes: ammonia-oxidizing bacteria (AOB), ammonia-oxidizing archaea (AOA), and nitrite-oxidizing bacteria (NOB). To date, no organism is known to be able to catalyze both autotrophic nitrification steps.

Anaerobic ammonium oxidation (the "anammox" process) was described as a shortcut in the biogeochemical N-cycle, where ammonium is oxidized, nitrite is reduced, and N_2 gas is formed (van de Graaf *et al.,* 1995). As anammox is catalyzed by strictly anaerobic and autotrophic bacteria, significant costs for aeration and carbon sources can be saved if the process is exploited for sewage treatment. Accordingly, a number of novel approaches for N-removal from wastewater combine partial nitrification (ammonia oxidation by AOB) with anammox (Third *et al.,* 2001; van Dongen *et al.,* 2001; Pynaert *et al.,* 2003; van der Star *et al.,* 2007).

All nitrifying microbes and anammox bacteria are very slow-growing organisms, which are sensitive to changes of their growth conditions. As a consequence, nitrification has often caused trouble in wastewater treatment due to unpredictable performance breakdowns. The causes of such events remain obscure in

 FISH Handbook for Biological Wastewater Treatment: Identification and quantification of microorganisms in activated sludge and biofilms by FISH. Edited by Per Halkjær Nielsen, Holger Daims and Hilde Lemmer. ISBN: 9781843392316. Published by IWA Publishing, London, UK.

many cases. Technical applications of anammox are young, and field reports are still rare, but the extremely slow growth of anammox bacteria may also become a problem when their populations decline for unknown reasons and the process fails in a full-scale WWTP. Therefore, fast methods to detect and quantify the involved organisms in activated sludge or biofilm are important tools not only for fundamental research, but could become relevant in the future also for the routine operation of nitrifying and anammox-based reactors: If a decline of any of the key populations is detected soon enough, operational countermeasures may be able to prevent a worst-case scenario. FISH certainly is the best available technique for this purpose, and a large set of rRNA-targeted probes is already available for the detection of AOB, NOB, and anammox bacteria. This chapter provides an overview of the most common nitrifiers and anammox organisms in WWTPs and of the FISH probes frequently used to detect and to quantify these microbes.

2.2 AMMONIA OXIDIZERS

In most nitrifying WWTPs, ammonia is oxidized by AOB of the genus *Nitrosomonas* (including *Nitrosococcus mobilis*). Most commonly found are AOB related to *N. europaea*, *N. eutropha*, *N. mobilis*, and *N. oligotropha* (Figure 2.1). Experiments using lab-scale reactors and studies of pure cultures suggested that *N. oligotropha* and closely related AOB are better adapted to low ammonia concentrations than *N. europaea* (Bollmann *et al.,* 2002, and references therein; Limpiyakorn *et al.,* 2007). However, detailed studies have been carried out only with a limited number of AOB strains. Furthermore, Lydmark *et al.* (2007) observed differential responses of *N. oligotropha*-related populations to ammonia shifts in a pilot-scale plant. Local differences in substrate concentrations (microniches) within flocs or biofilms may also affect the distribution and activity of AOB in WWTPs (Gieseke *et al.,* 2005). *Nitrosococcus mobilis*-related AOB seem to occur especially in reactors treating sludge liquor or other types of wastewater with elevated ammonia and salt concentrations such as animal rendering waste (Juretschko *et al.,* 1998).

In activated sludge flocs and biofilms, AOB related to *Nitrosomonas* usually form almost spherical, compact cell aggregates (Figure 9.1, Chapter 9). Single cells within these clusters are well visible at 630× or 1000× magnification. The diameter of most AOB cell clusters is 10–50 μm. Less compact aggregates occur occasionally, where the cells are more irregularly arranged with more space between them. Single AOB cells are seldom found in activated sludge by microscopy, but they may easily be overlooked when dense floc or biofilm structures are observed.

AOB of the genus *Nitrosospira* have occasionally been detected in WWTPs, but these AOB are generally more common in terrestrial habitats and seem to play only minor roles for wastewater treatment. Notable exceptions are rhizoremediation plants, where *Nitrosospira*-related organisms seem to be more frequent (Haleem *et al.,* 2000).

Both genera *Nitrosomonas* and *Nitrosospira* belong to the *Betaproteobacteria*. In marine habitats, *Nitrosomonas marina* and the gammaproteobacterial lineage of AOB with the two species *Nitrosococcus oceani* and *N. halophilus* are common. Due to their high salt requirements, these AOB are absent from most WWTPs with the possible exception of brackish water and saltwater systems: *N. marina* was found in a nitrifying trickling filter biofilm associated with a marine aquaculture system (Foesel *et al.,* 2008). Occurrence of *N. oceani* and *N. halophilus* in WWTPs has not been reported yet.

The question of whether AOA might be important for nitrification in WWTPs still is a matter of active research and dispute. Indeed, AOA have been detected in activated sludges by means of specific polymerase chain reaction (PCR) (Park *et al.,* 2006). However, PCR does not allow to distinguish intact, active cells from naked DNA or metabolically inactive cells that are occasionally swept into the activated

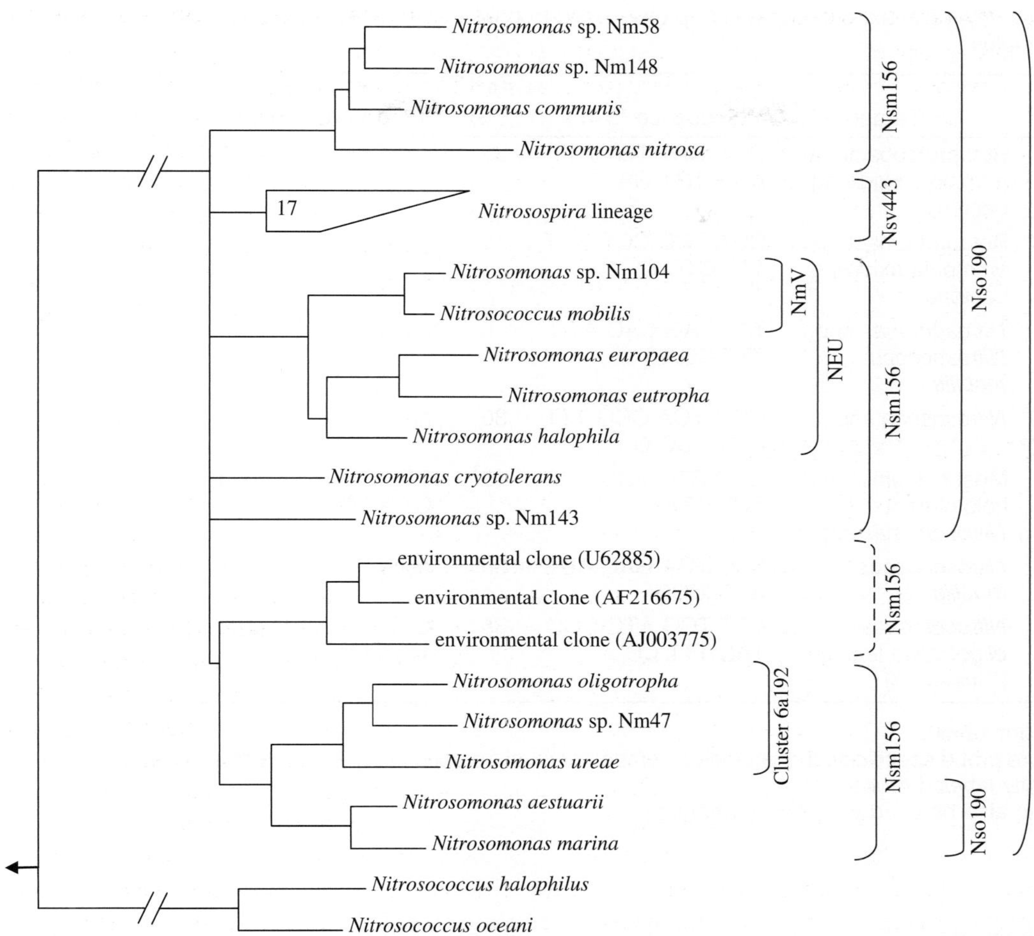

Figure 2.1. 16S rRNA-based phylogenetic tree showing main lineages of known bacterial AOB. Brackets indicate the coverage of the probes listed in Table 2.1. Dashed brackets indicate incomplete sequence information at the respective probe target site, so that precise coverage information cannot be inferred. The exact branching order of the AOB lineages cannot yet be determined due to a limited phylogenetic resolution of the 16S rRNA marker gene. Therefore, unclear tree topology is depicted as multifurcation.

sludge systems but have no relevant functions there. So far, high abundance of AOA in nitrifying WWTPs has not been confirmed by more robust methods such as quantitative FISH with rRNA-targeted probes.

2.2.1 Probes for the detection of AOB

The oligonucleotide probes, which have most frequently been used to detect and quantify AOB in WWTPs, are listed in Table 2.1. The tree in Figure 2.1 illustrates the specificity and coverage of the various probes. Please note that this compilation of probes is not complete (i.e. additional AOB-targeted probes have been published). However, the probes listed here have been tested in a large number of studies by different research groups. Their coverage and specificity have been re-evaluated, based on increasingly large rRNA sequence databases, since the probes were originally published.

Table 2.1. rRNA-targeted oligonucleotide probes used to detect AOB in nitrifying activated sludge and biofilm samples.

Probe name	Target	Sequence (5′-3′)	FA[1)] (%)	Competitor oligonucleotide[2)]	Reference
Nso1225	Betaproteobacterial ammonia-oxidizing bacteria	CGC CAT TGT ATT ACG TGT GA	35	none	Mobarry *et al.* 1996
Nso190	Betaproteobacterial ammonia-oxidizing bacteria	CGA TCC CCT GCT TTT CTC C	55	none	Mobarry *et al.* 1996
Nsm156	*Nitrosomonas* spp., *Nitrosococcus mobilis*	TAT TAG CAC ATC TTT CGA T	5	none	Mobarry *et al.* 1996
Nsv443	*Nitrosospira* spp.	CCG TGA CCG TTT CGT TCC G	30	none	Mobarry *et al.* 1996
NEU	Most halophilic and halotolerant *Nitrosomonas* spp.	CCC CTC TGC TGC ACT CTA	40[3)]	TTC CAT CCC CCT CTG CCG	Wagner *et al.* 1995
NmV	*Nitrosococcus mobilis*	TCC TCA GAG ACT ACG CGG	35	none	Juretschko *et al.* 1998
Cluster 6a192	*Nitrosomonas oligotropha* lineage (Cluster 6a)	CTT TCG ATC CCC TAC TTT CC	35	CTT TCG ATC CCC TGC TTC C	Adamczyk *et al.* 2003

1) FA = Formamide.
2) To ensure probe specificity, the unlabeled competitor must be used in equimolar amounts together with the fluorescently labeled probe.
3) NEU can also be used with 35% formamide.

If the aim is to detect and quantify most of the AOB in an activated sludge sample at once, it is convenient to use a probe mixture consisting of the probes Nso1225 + NEU + NmV + Cluster6a192 in equimolar amounts (and also the respective competitor oligonucleotides). All probes except the competitors should then be labeled with the same fluorochrome. All the probes in this "AOB-mix" can be used with 35% formamide in the hybridization buffer. In order to differentiate the AOB in a sample, the same "AOB-mix" can be used, but then with probes labeled with different fluorochromes. In addition, one could apply the other probes (Table 2.1) using the respective formamide concentrations.

For unknown reasons, probe Nso1225 occasionally yields only dim fluorescence signals even though the target AOB are present in the sample and their cellular ribosome content would be high enough for a bright fluorescence signal after FISH. Additional test hybridizations with probe Nso190 should be carried out to ensure that no AOB are overlooked in a sample. The coverage of Nso190 resembles that of Nso1225, but Nso190 is more reliable in terms of probe brightness. However, due to its requirement for 55% formamide, Nso190 cannot be mixed with the other probes and has to be applied separately.

Please note that probe Nsv443 seldom yields unambiguously positive signals when applied to activated sludge. In some cases a few cells are stained by this probe in each field of view, but their density seems too low for technical relevance. Whether such cells are *Nitrosospira* spp. or unspecifically detected non-target organisms remains to be shown. As mentioned already, higher densities of *Nitrosospira* spp. have not frequently been observed in full-scale wastewater treatment plants. If cells are detected after FISH with Nsv443 in activated sludge, it is recommended that additional control experiments should be performed to check for autofluorescent cells, or unspecific binding of DNA or of the fluorochromes to cell surfaces. This can be accomplished by (i) using a nonsense probe, which cannot target any bacterium; (ii) using the same probe labeled with a different fluorochrome; or (iii) adding large amounts of DNA (e.g. salmon sperm DNA) to the sample prior to FISH in order to saturate any unspecific DNA binding sites on cell surfaces. Furthermore, a hierarchy of phylogenetically nested probes should be used to confirm the identity of probe-positive cells. In case of *Nitrosospira* this could be a combination of Nsv443 with probes Nso1225 or Nso190, respectively, which should be labeled with a different fluorochrome than Nsv443.

2.3 NITRITE OXIDIZERS

In the majority of nitrifying WWTPs, the dominant NOB are members of the genus *Nitrospira* (e.g. Juretschko *et al.,* 1998; Schramm *et al.,* 1998; Daims *et al.,* 2001a; Gieseke *et al.,* 2003). This genus is part of the distinct bacterial phylum *Nitrospirae*, and thus is not closely related to the other known NOB, which are all *Proteobacteria* (Figure 2.2). All *Nitrospira* are slow-growing and recalcitrant bacteria, which are very difficult to culture in the laboratory. Only three isolates have been obtained from other sources than activated sludge, whereas no *Nitrospira* has been isolated from any WWTP although one high enrichment was achieved (Spieck *et al.,* 2006). Therefore, the discovery that *Nitrospira* (not *Nitrobacter*) are the key NOB in full-scale wastewater treatment plants (Juretschko *et al.,* 1998) was based on the use of FISH and other cultivation-independent methods. *Nitrospira* is a diverse genus that consists of several phylogenetic sublineages (Figure 2.2). In WWTPs, the sublineages I, II and IV are found, but in many systems sublineage I *Nitrospira* are predominant. Evidence exists that sublineage I can outcompete sublineage II *Nitrospira* at (transiently) elevated nitrite concentrations (Maixner *et al.,* 2006). Sublineage IV, which occurs in nature in marine habitats, is limited to WWTPs receiving sewage with high salt concentrations.

All known *Nitrospira* in WWTPs form spherical or irregularly shaped cell aggregates, which consist of several hundred or thousand cells (Figure 9.2, Chapter 9). The diameter of these aggregates is 10–100 μm, but even larger clusters are found occasionally. In particular the large *Nitrospira* aggregates often contain narrow "channels" and larger cavities (Figure 9.2, Chapter 9).

In contrast to *Nitrospira*, the genus *Nitrobacter* seems to play a minor role in wastewater treatment. *Nitrobacter* cells occur in many reactors and can be enriched or isolated by incubation of activated sludge in artificial nitrite media, but FISH has shown that *Nitrobacter* do usually not reach significant cell densities in WWTPs (Wagner *et al.,* 1996). Notable exceptions are reactors that temporarily contain elevated nitrite concentrations, for example sequencing batch reactors treating highly concentrated sludge liquor. In such systems, *Nitrobacter* can also reach a relatively high abundance (Daims *et al.,* 2001a), probably because these NOB are adapted to higher NO_2^- concentrations whereas *Nitrospira* are adapted to lower NO_2^- concentrations (Schramm *et al.,* 1999). *Nitrobacter* in WWTPs form cell aggregates like the other nitrifiers, or they occur as single cells embedded in the biofilm matrix.

In flocs and biofilms, NOB often occur in the direct spatial neighborhood to AOB, which reflects the mutualistic symbiosis of these two functional groups (Maixner *et al.,* 2006) (Figure 9.3, Chapter 9).

Other NOB comprise the marine genera *Nitrococcus* and *Nitrospina* and the only recently discovered betaproteobacterial *Nitrotoga arctica* (Alawi *et al.,* 2007). To date, none of these NOB has been shown, by FISH or any other quantitative cultivation-independent assay, to be functionally important for nitrification in wastewater treatment.

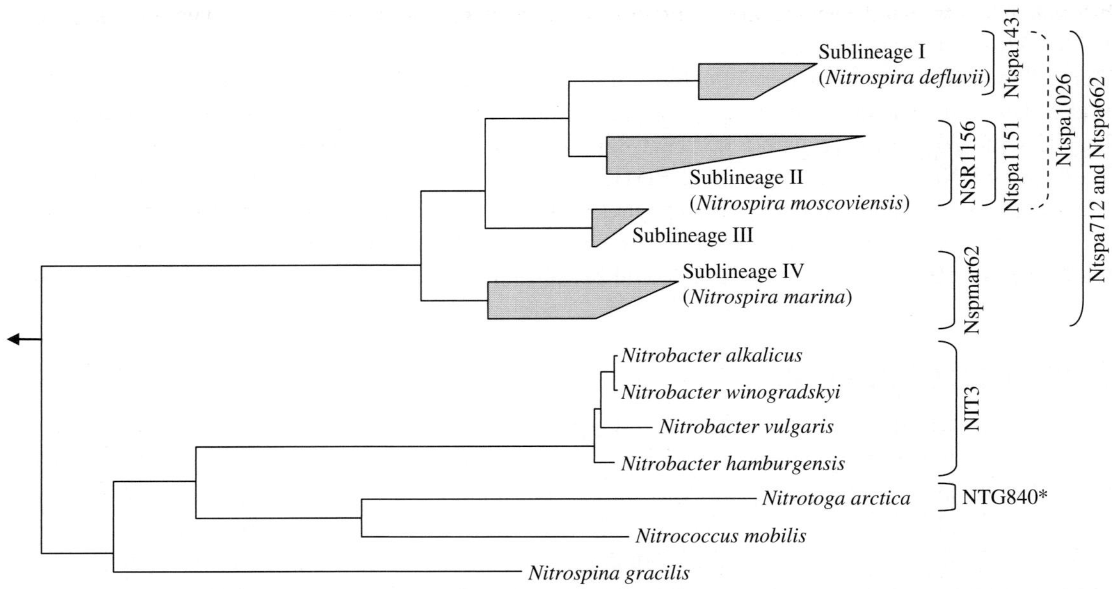

Figure 2.2. 16S rRNA-based phylogenetic tree showing main lineages of known bacterial NOB. Brackets indicate the coverage of the probes listed in Table 2.2. Dashed brackets indicate incomplete coverage of the target groups. Sublineages I–IV comprise the genus *Nitrospira*. (∗) Probe NTG840 may also bind to some nontarget *Betaproteobacteria*, which have no 16S rRNA sequence mismatch at the probe binding site.

2.3.1 Probes for the detection of NOB

The oligonucleotide probes, which have most frequently been used to detect and quantify NOB in WWTPs, are listed in Table 2.2. The tree in Figure 2.2 illustrates the specificity and coverage of the various probes. Please note that this compilation of probes is not complete (i.e. additional NOB-targeted probes have been published). However, most of the probes listed here have been tested in a large number of studies by different research groups. Their coverage and specificity have been re-evaluated, based on increasingly large rRNA sequence databases, since the probes were originally published.

In most cases, probes Ntspa662 targeting the genus *Nitrospira* and NIT3 targeting *Nitrobacter* are sufficient to detect NOB in WWTPs. Probe Ntspa662 can be combined with Ntspa712, which also targets all *Nitrospira*, or with Ntspa1431, Ntspa1151 and Nspmar62, which target different sublineages of this genus. The latter combinations are especially interesting if the probes are labeled with different fluorescent dyes, so that one can determine whether different *Nitrospira* co-exist in the same WWTP.

Probes Ntspa1026 and Nsr1156 usually yield excellent signal intensities with targeted *Nitrospira*, but neither probe completely covers the respective *Nitrospira* sublineages that are relevant for wastewater treatment. Therefore, the use of these probes is recommended only in combination with either Ntspa662 or Ntspa712 to increase the likelihood that all *Nitrospira* in a sample are detected.

2.4 ANAMMOX BACTERIA

All known anaerobic ammonium oxidizers belong to the phylum *Planctomycetes* (Strous *et al.,* 1999). They are peculiar prokaryotes in many respects, but their most distinct feature is intracellular

Table 2.2. rRNA-targeted oligonucleotide probes used to detect NOB in nitrifying activated sludge and biofilm samples.

Probe name	Target	Sequence (5′-3′)	FA[1] (%)	Competitor oligonucleotide[2]	Reference
Ntspa712	Phylum *Nitrospirae*	CGC CTT CGC CAC CGG CCT TCC	50[3]	CGC CTT CGC CAC CGG TGT TCC	Daims *et al.* 2001a
Ntspa662	Genus *Nitrospira*	GGA ATT CCG CGC TCC TCT	35	GGA ATT CCG CTC TCC TCT	Daims *et al.* 2001a
Ntspa1026	*Nitrospira* sublineages I and II[4]	AGC ACG CTG GTA TTG CTA	20	none	Juretschko *et al.* 1998
Ntspa1431	*Nitrospira* sublineage I	TTG GCT TGG GCG ACT TCA	35	none	Maixner *et al.* 2006
Ntspa1151	*Nitrospira* sublineage II	TTC TCC TGG GCA GTC TCT CC	35	none	Maixner *et al.* 2006
Nsr1156	*Nitrospira* sublineage II	CCC GTT CTC CTG GGC AGT	30	none	Schramm *et al.* 1998
Nspmar62	*Nitrospira marina*-related *Nitrospira*	GCC CCG GAT TCT CGT TCG	40	none	Foesel *et al.* 2008
NIT3	Genus *Nitrobacter*	CCT GTG CTC CAT GCT CCG	40	CCT GTG CTC CAG GCT CCG	Wagner *et al.* 1996
NTG840	*Nitrotoga arctica*[5]	CTA AGG AAG TCT CCT CCC	10–20	none	Alawi *et al.* 2007

[1] FA = Formamide.
[2] To ensure probe specificity, the unlabeled competitor must be used in equimolar amounts together with the fluorescently labeled probe.
[3] Ntspa712 can also be used with 35% formamide, especially if combined with Ntspa662.
[4] Ntspa1026 does not cover all members of these *Nitrospira* sublineages.
[5] NTG840 has not yet been used with activated sludge. Nevertheless it is listed here, because *Nitrotoga*-like bacteria have been enriched from a WWTP (Alawi *et al.,* 2007) and NTG840 is the only published probe targeting this organism. Please note that NTG840 is not fully specific as some nontarget bacteria have no 16S rRNA sequence mismatches at the probe binding site.

compartmentalization, which is very uncommon in bacteria (Lindsay *et al.,* 2001). Since the discovery of anammox bacteria about a decade ago, several different representatives have been enriched from WWTPs. These are the lineages *Brocadia* with the candidate species "B. anammoxidans" (Strous *et al.,* 1999) and "B. fulgida" (Kartal *et al.,* 2008), *Kuenenia* with "K. stuttgartiensis" (Schmid *et al.,* 2000), *Scalindua* with "S. brodae" and "S. wagneri" (Schmid *et al.,* 2003), and *Anammoxoglobus* with "A. propionicus" (Kartal *et al.,* 2007). The best studied anammox organism so far is *K. stuttgartiensis* whose genome was sequenced, by environmental genomics, and was used to reconstruct the intricate anammox biochemical pathway (Strous *et al.,* 2006). Interestingly, *A. propionicus* co-oxidizes propionate and ammonium (Kartal *et al.,* 2007), whereas *B. fulgida* co-oxidizes acetate and ammonium more efficiently than the other anammox bacteria (Kartal *et al.,* 2008). These physiological differences suggest adaptations of the various anammox organisms to specific ecological niches, which could affect the anammox community composition in WWTPs.

Table 2.3. rRNA-targeted oligonucleotide probes used to detect anammox organisms in activated sludge and biofilm samples.

Probe name	Target	Sequence (5′-3′)	FA[1] (%)	Competitor oligonucleotide	Reference
Pla46	all *Planctomycetes*	GAC TTG CAT GCC TAA TCC	30	none	Neef *et al.* 1998
Amx368	all anammox bacteria	CCT TTC GGG CAT TGC GAA	15	none	Schmid *et al.* 2003
Amx820	*B. anammoxidans* *K. stuttgartiensis*	AAA ACC CCT CTA CTT AGT GCC C	40	none	Schmid *et al.* 2000
Kst157	*K. stuttgartiensis*	GTT CCG ATT GCT CGA AAC	25	none	Schmid *et al.* 2001
Amx1015	*B. anammoxidans*	GAT ACC GTT CGT CGC CCT	60	none	Schmid *et al.* 2000
Bfu613	*B. fulgida*	GGA TGC CGT TCT TCC GTT AAG CGG	30	none	Kartal *et al.* 2008
Apr820	*A. propionicus*	AAA CCC CTC TAC CGA GTG CCC	40	none	Kartal *et al.* 2007
BS820	*S. wagneri*	TAA TTC CCT CTA CTT AGT GCC C	40	none	Kuypers *et al.* 2003
Scabr1114	*S. brodae*	CCC GCT GGT AAC TAA AAA CAA G	20	none	Schmid *et al.* 2003

[1] FA = Formamide.

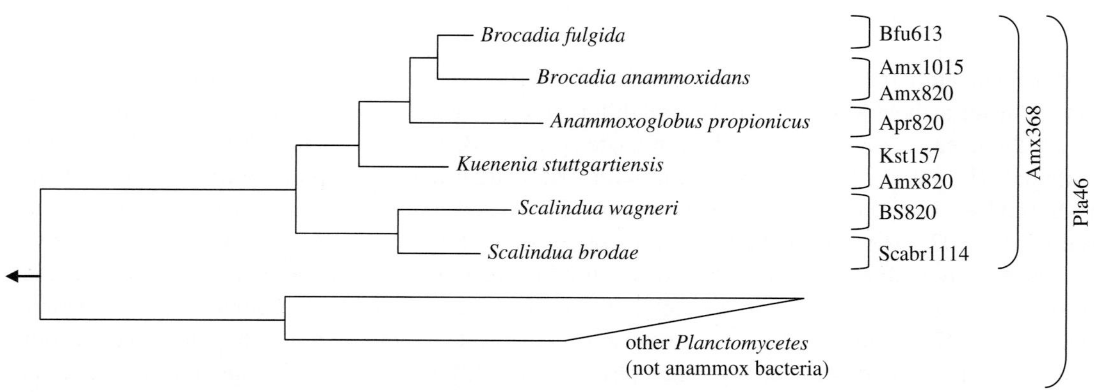

Figure 2.3. 16S rRNA-based phylogenetic tree showing main lineages of known anammox organisms. Brackets indicate the coverage of the probes listed in Table 2.3.

Usually, anammox bacteria occur as small to large cell aggregates (Figure 9.4, Chapter 9). Anammox bacteria have a coccoid morphology and a 'donut-like' shape after FISH with an apparent small hole in the centre of the cells. This 'hole' is caused by the absence of ribosomes (and thus, of target molecules for FISH probes) in this region. Hence, in combination with suitable FISH probes, anammox organisms are easy to identify by microscopy. However, one should keep in mind that other planctomycetes, which are morphologically similar, also occur in WWTPs (Neef *et al.,* 1998). These planctomycetes probably are heterotrophs not involved in the anammox process, and they are also labeled by general planctomycetes-specific FISH probes such as Pla46. Thus, anammox-specific FISH probes (Schmid *et al.,* 2005) should always be used to confirm that donut-shaped cells resembling planctomycetes really are anammox bacteria.

2.4.1 Probes for the detection of anammox organisms

The oligonucleotide probes, which are best for detecting and quantifying anammox bacteria in WWTPs, are listed in Table 2.3. The tree in Figure 2.3 illustrates the specificity and coverage of the various probes. Note that some additional probes targeting anammox bacteria have been published (a compilation can be found in Schmid *et al.,* 2005). However, these other probes do not offer a better coverage and specificity than the probes listed in Table 2.3 for anammox organisms occurring in full-scale WWTPs.

3

Identification of denitrifying microorganisms in activated sludge by FISH

Jeppe Lund Nielsen and Aviaja A. Hansen

3.1 INTRODUCTION

Denitrification is defined as the anoxic process in which nitrate or nitrite is reduced to gaseous nitrogen oxides (nitrogen oxide, nitrous oxide or free nitrogen). The process has been the focus of numerous studies due to its major importance in the global N-cycle, for nitrogen removal in the wastewater industry, and lately also for potential production of the greenhouse gas N_2O. Although subject for many research studies, relatively little is known about the microorganisms carrying out the process in full-scale WWTPs and in particular their phylogenetic distribution (Juretschko *et al.,* 2002; Wagner and Loy, 2002). However, a few recent studies have revealed some important denitrifiers in full-scale WWTPs (Thomsen *et al.,* 2007; Hagmann *et al.,* 2008; Morgan-Sagastume *et al.,* 2008).

The denitrifying organisms are facultative anaerobic heterotrophs and, less frequently, autotrophs. The interaction of the denitrification process and the nitrification process is the main cause for removal of nitrogen from wastewater. While nitrification primarily occurs under aerobic conditions, denitrification requires anoxic conditions with the presence of nitrate or nitrite and easily degradable organic substrates. Under oxic conditions most denitrifiers carry out full aerobic respiration. These facultative denitrifiers seem to constitute a major fraction of the whole of bacteria in the activated sludge community (Nielsen and Nielsen, 2002a; Morgan-Sagastume *et al.,* 2008). Investigations on the denitrification rates in the presence of various electron donors reveal the presence of several specialized populations of denitrifiers in activated sludge (Thomsen *et al.,* 2007; Hagmann *et al.,* 2008; Morgan-Sagastume *et al.,* 2008). Some

 FISH Handbook for Biological Wastewater Treatment: Identification and quantification of microorganisms in activated sludge and biofilms by FISH. Edited by Per Halkjær Nielsen, Holger Daims and Hilde Lemmer. ISBN: 9781843392316. Published by IWA Publishing, London, UK.

microorganisms are only partially involved in the denitrification process for example by reducing nitrate and/or nitrite to nitric oxide or nitrous oxide but these are not well described.

3.2 IDENTITY OF DENITRIFIERS IN WASTEWATER TREATMENT SYSTEMS

The organisms capable of denitrification are not limited to specific microbial taxa but present in many phylogenetic subgroups. Denitrifiers are found among the *Proteobacteria*, *Firmicutes*, and the *Bacteroidetes* covering more than 50 genera (Zumft, 1992). Previously, culture-dependent studies of denitrifiers typically described these to be members of *Pseudomonas*, *Bacillus* or *Alcaligenes*, but these are usually not the dominant genera in the activated sludge process. Several publications based on both culture-dependent and culture-independent approaches indicate that other denitrifiers must be dominating the activated sludge system (e.g. Heylen *et al.*, 2006a,b; Hagman *et al.*, 2008).

Culture-independent studies carried out directly in a WWTP treating industrial wastewater (e.g. Juretschko *et al.*, 2002; Wagner and Loy, 2002) have shown the betaproteobacterial genus *Azoarcus* (*Rhodocyclaceae*) as the dominating denitrifier, while similar studies in treatment plants treating municipal wastewater have revealed that several betaproteobacterial genera belonging to the families *Comamonadaceae* and *Rhodocyclaceae* are present. Within the *Comamonadaceae*, *Curvibacter*-related organisms (reclassified from *Aquaspirillum*, Ding and Yokota, 2004) and within the *Rhodocyclaceae* members of the genera *Zoogloea, Azoarcus*, and *Thauera* are the dominating denitrifiers (Rossellò-Mora *et al.,* 1995; Juretschko *et al.,* 2002; Thomsen *et al.,* 2004; 2007). In WWTPs with Enhanced Biological Phosphorus Removal (EBPR) with combined nitrogen removal, polyphosphate accumulating members of *Accumulibacter* (*Betaproteobacteria*) are present and some of these, but presumably not all, are capable of denitrification (e.g. Kong *et al.,* 2004). The identity of these subgroups of denitrifying *Accumulibacter* is not well known and it has hitherto not been possible to phylogenetically distinguish these by application of gene probes or by other molecular methods (Seviour *et al.,* 2003). These groups of *Betaproteobacteria* constitute the vast majority of denitrifying microorganisms in most WWTPs (Thomsen *et al.,* 2007), although several other, less abundant unclassified denitrifiers also are present (Morgan-Sagastume *et al.,* 2008). Culture-dependent studies have shown that denitrifying organisms belonging to the alphaproteobacterial genera *Paracoccus (Rhodobacteraceae)* and *Rhizobium (Rhizobiaceae)* are present in WWTPs (Heylen *et al.,* 2006a,b). However, due to the lack of appropriate probes the abundance of these denitrifying *Alphaproteobacteria* has not been further evaluated.

Culture-dependent studies of wastewater treatment systems fed with methanol under denitrifying conditions frequently identify *Alphaproteobacteria* belonging to the genus *Hyphomicrobium* (*Hyphomicrobiaceae*) (e.g. Timmermans and Van Haute 1983), which have been further verified by dot blot analysis on the activated sludge (Layton *et al.,* 2000).

Studies carried out in denitrifying lab-scale reactors inoculated with activated sludge and run on acetate or other single substrates reveal high diversities of potential denitrifying organisms belonging to *Proteobacteria, Bacteroidetes, Chloroflexi, Firmicutes*, and *Actinobacteria* (Yosie *et al.,* 2001; Ginige *et al.,* 2004; Osaka *et al.,* 2006). This may not, however, reflect the indigenous consortia abundant in full-scale WWTPs.

3.3 ABUNDANT DENITRIFIERS IN FULL-SCALE PLANTS

Family *Comamonadaceae:*

Genus *Curvibacter*. *Curvibacter* has been found to be a potential dominating denitrifier in full-scale plants without external carbon addition with abundances of 11–29% of all *Bacteria* (Thomsen *et al.,* 2004).

Its morphology is quite distinct and easy to recognize. It is typically large and coccoid with a diameter of 1–1.5 μm and grows as single cells or in microcolonies typically ranging from 10–15 μm in diameter, see Figure 9.5, Chapter 9.

Family *Rhodocyclaceae:*
Genus *Azoarcus*. The abundance of *Azoarcus* in some full-scale plants varies from 3–16% of the total biomass (Thomsen *et al.,* 2007), but in plants with addition of an external carbon source in the form of methanol more than 30% have been found (Hagman *et al.,* 2008). Different morphotypes have been observed, but usually they appear rod-shaped (0.5–1 μm × 1–2 μm) and grow as single cells or in relatively small microcolonies (5–20 μm in diameter), see Figure 9.6, Chapter 9.

Genus *Thauera*. The number of *Thauera*-related bacteria ranges from 2–11% of the total biomass in some full-scale plants (Thomsen *et al.,* 2007). The cells usually grow in microcolonies of various sizes (5–40 μm), and the rod-shaped cell size is typically 1.0 × 2.5 μm, see Figure 9.7, Chapter 9.

Genus *Zoogloea*. *Zoogloea*-related bacteria are typically observed in small numbers except for a few plants in which up to 10% of the total biomass has been reported (Rossellò-Mora *et al.,* 1995). *Zoogloea*-related bacteria have often been described as forming colonies in typical branched gelatinous matrices, the so-called *Zoogloea* fingers. Most cells are rod-shaped (1–1.5 μm × 1 μm).

Genus *Accumulibacter*. The presence of the potential PAO *Accumulibacter* usually indicates the presence of an EBPR configuration with abundances ranging from 3–15% of the total biomass. The cells are typically rod-shaped (1–1.5 μm) and almost always growing in microcolonies (Larsen *et al.,* 2006). See also Chapter 4 about PAOs.

Other denitrifiers, such as *Acidovorax,* are usually observed but only in small numbers, typically less than 1–2% of the total biomass (Heylen *et al.,* 2008). Glycogen-accumulating organisms (e.g. *Competibacter*) contain members that are capable of denitrification (Zeng *et al.,* 2003), see Chapter 4.

Other potential denitrifiers
Denitrifiers from several genera have been isolated from activated sludge. Figure 3.1 shows a selection of the phylogenetically identified isolates. The tree has also been supplemented with sequences from three denitrifying clone libraries, derived i) from stable isotope probing of denitrifiers grown in lab-scale reactors on acetate and methanol (Ginige *et al.,* 2004, 2005; and Osaka *et al.,* 2006) and ii) denitrifying pure cultures from lab-scale reactors seeded with wastewater (Etchebehere *et al.,* 2001; Yosie *et al.,* 2001; Martin *et al.,* 2006; Wang *et al.,* 2007). Recently, it was found that hitherto unidentified bacteria with a potential denitrifying capability (besides from the *Betaprotobacteria*) may be found within the groups of *Alphaproteobacteria*, *Gammaproteobacteria*, and *Actinobacteria*. The study revealed that their relative abundance in the activated sludge investigated was almost as important as the *Betaproteobacteria* (Morgan-Sagastume *et al.,* 2008).

3.4 PROBES FOR DETECTION OF DENITRIFIERS

In Table 3.1 the oligonucleotide probes that can be used for identification and quantification of denitrifiers or potential denitrifiers are listed. In the phylogenetic tree in Figure 3.1 the specificities and coverage of the probes can be seen. At present it is not possible to target all denitrifiers, but for detection of the majority of the *Betaproteobacteria* it is recommended to use the following probes:

Of the *Comamonadaceae* family, most *Acidovorax* can be targeted by probe ACI208 (Amann *et al.,* 1996), but also *Variovorax* are targeted. Some *Curvibacter* are targeted by Curvi997 with a high specificity (Thomsen *et al.,* 2004).

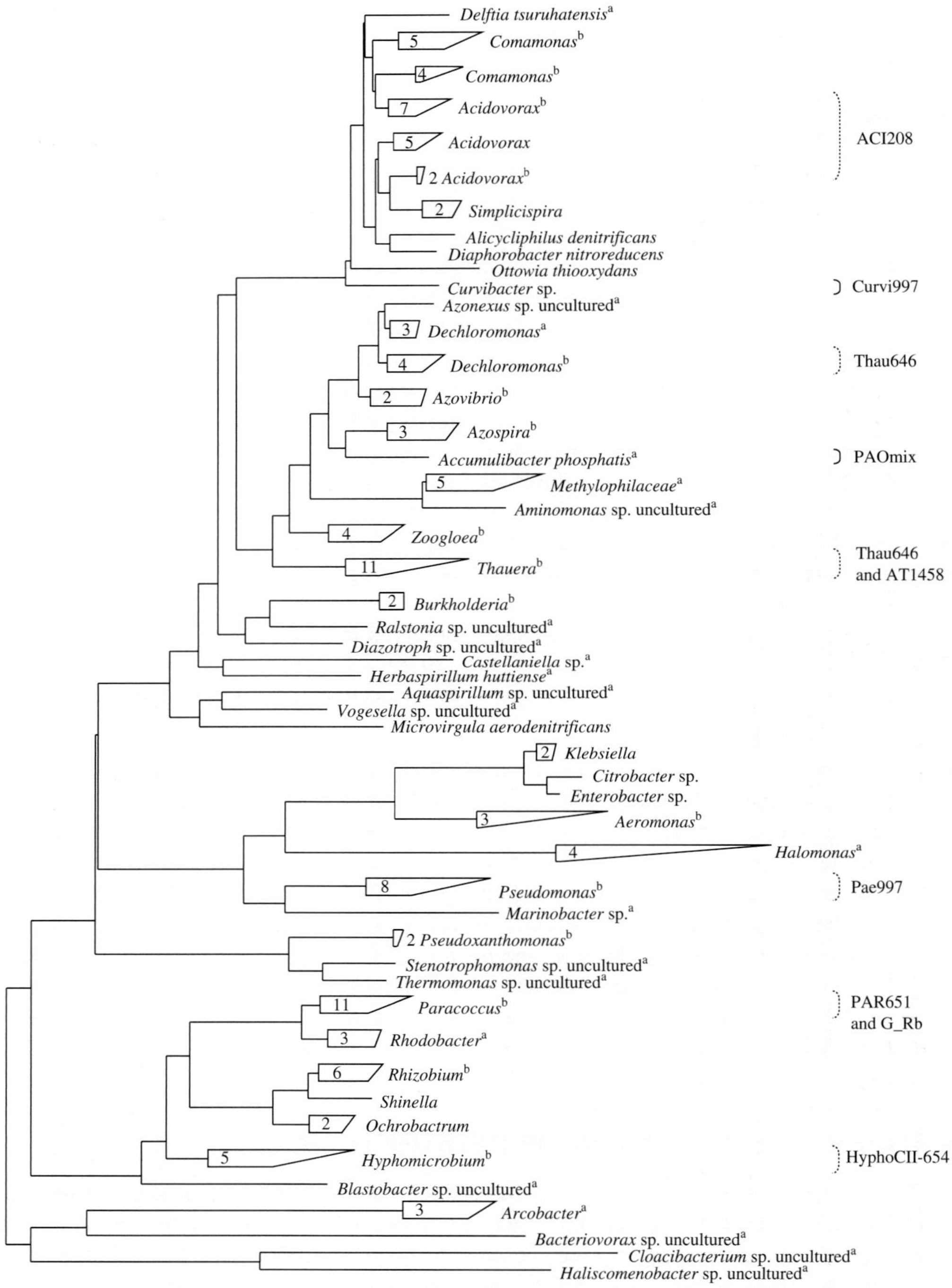

Figure 3.1. 16S rRNA-based phylogenetic tree showing denitrifying bacteria from activated sludge. Brackets indicate the coverage of the probes listed in Table 3.1. Dashed brackets indicate incomplete coverage of the indicated groups. [a]Sequences from bioreactor studies, [b]Sequences from full-scale and bioreactor studies.

Table 3.1. 16S rRNA oligonucleotide probes for the identification of potential denitrifiers.

Probe name	Target	Sequence 5′-3′	FA conc	Competitor oligonucleotide	Reference
PAR 651	Genus *Paracoccus*	ACC TCT CTC GAA CTC CAG	40	None	Neef *et al.*, 1996
G_Rb	*Rhodobacter, Roseobacter*	GTC AGT ATC GAG CCA GTG AG	30	None	Giuliano *et al.*, 1999
HyphoCII-654	*Hyphomicrobium denitrificans, H. methylovorum, H. facilis*	CCC ACC TCT ATC GGA CTC	10	None	Layton *et al.*, 2000
Curvi997*	*Curvibacter*	CTC TGG TAA CTT CCG TAC	35	2 competitor probes CTC TGG CAA CTT CCG TAC CTC TGG TCA CTT CCG TAC	Thomsen *et al.*, 2004
PAOmix	Most *Accumulibacter*	PAO462, PAO651 and PAO846	35	None	Crocetti *et al.*, 2000
AZA645**	Most members of the *Azoarcus* cluster	GCC GTA CTC TAG CCG TGC	20	None	Hess *et al.*, 1997
Thau646	*Thauera*	TCT GCC GTA CTC TAG CCT T	45		Lajoie *et al.*, 2000
ACI208	*Acidovorax* spp.	CGC GCA AGG CCT TGC	20	None	Amann *et al.*, 1996
ZRA23a	Most members of the *Zoogloea* lineage, not *Z. resiniphila*	CTG CCG TAC TCT AGT TAT	35	None	Rosselló-Mora *et al.*, 1995
AT1458	*Azoarcus-Thauera* cluster	GAA TCT CAC CGT GGT AAG CGC	50	None	Rabus *et al.*, 1999
Pae997	most true *Pseudomonas* spp.	GCT GGC CTA GCC TTC	0	None	Amann *et al.*, 1996

* Renamed from Aqs997
** Renamed from AZO644

Of the family *Rhodocyclaceae,* most *Azoarcus* can be targeted by probe AZA645 (Hess *et al.*, 1997). Probe Thau646 (formerly probe MZ1) targets *Thauera* (Lajoie *et al.,* 2000) but with a low specificity and with a high proportion of outgroup hits including *Dechloromonas* of the *Rhodocyclaceae* family, *Comamonas* of the *Comamonadaceae* family and *Aquabacterium* of the Incertae Sedis 5 (Figure 3.1). Most *Azoarcus* and *Thauera* can be detected by probe AT1458 (Rabus *et al.,* 1999). The probe is very specific, apart from targeting most of the *Castellaniella* of the *Alcaligenaceae* family. ZRA (identical to ZRA23a) targets a high proportion of *Zoogloea*-related organisms (Rosselló-Mora *et al.,* 1995), but not all known *Zoogloea* strains from WWTPs (Figure 3.1).

Accumulibacter-related PAOs can be targeted by the PAOmix (Crocetti *et al.,* 2000), but it targets also a high proportion of *Propionivibrio* of the *Rhodocyclaceae* family.

Specific probes for the denitrifying *Alphaproteobacteria* and *Gammaproteobacteria* have yet to be developed and therefore the probes available have a low coverage of the denitrifiers in WWTPs (Figure 3.1). The following probes can be used to target *Alphaproteobacteria*:

PAR651 targets most *Paracoccus* of the family *Rhodobacteraceae* with a high specificity (Neef *et al.,* 1996). The denitrifying *Hyphomicrobium* is specifically targeted by probe HyphoCII-654 (Layton *et al.*, 2000) (see Figure 9.9, Chapter 9). G_Rb is a broad probe targeting most *Rhodobacteraceae* and some *Phyllobacteriaceae,* see Figure 9.10, Chapter 9. (Giuliano *et al.*, 1999; Eilers *et al.,* 2000) and has therefore also many non-denitrifying organism targets.

The following probe can be used to target some potential denitrifiers within the *Gammaproteobacteria*: Pae997 targets *Pseudomonas* (Amann *et al.,* 1996) with a high specificity, see Figure 9.11, Chapter 9.

4

Identification of polyphosphate-accumulating and glycogen-accumulating organisms by FISH

Per Halkjær Nielsen, Hien Thi Thu Nguyen, Simon Jon McIlroy, Artur Tomasz Mielczarek, and Robert Seviour

4.1 INTRODUCTION

Enhanced biological phosphorus removal (EBPR) processes have been widely used to remove soluble phosphate from domestic wastewater. Consistent and reliable operation of these systems requires successful selection and enrichment of microorganisms that biologically accumulate Pi into granules of intracellular polyphosphate.

The microbial communities in EBPR processes comprise populations considered important for EBPR [i.e. polyphosphate-accumulating organisms (PAOs)], those competing with PAOs [i.e. glycogen-accumulating organisms (GAOs)], and yet-to-be identified populations whose functions and interactions with the PAOs and GAOs remain to be further understood. PAOs in full-scale systems are mainly represented by the betaproteobacterial *Candidatus* Accumulibacter phosphatis (hereafter called *Accumulibacter*) or related species that consist of members with and without an ability for denitrification. Proliferation of glycogen-accumulating organism populations can cause failure in EBPR plants as they compete with the PAOs for organic substrates during the anaerobic period without contributing to Pi removal (Oehmen *et al.*, 2007). Originally they were referred to as the "G bacteria" (Cech and Hartman, 1990) and more recently as the tetrad-forming microorganisms (TFOs) morphotype, since some appear as distinctive coccal-shaped cells growing in tetrads (Tsai and Liu, 2002). The term glycogen-accumulating organisms (GAOs) was first suggested by Mino *et al.* (1995) defining a phenotype that stores glycogen aerobically and consumes it anaerobically for energy production for the uptake of short-chain fatty acids for PHA production. However, not all GAOs appear in tetrads. It has been reported that in the absence of PAOs, phylogenetically different GAOs can compete with each other in lab-scale reactors. Less is known about their competition in full-scale plants.

A comprehensive set of rRNA-targeted oligonucleotide probes has been applied to identify *in situ* such populations in full-scale EBPR systems around the world including countries like Japan, USA, Denmark, and Australia. The role of *Accumulibacter*-related PAOs seems less important in full-scale EBPR processes compared to laboratory-scale EBPR systems fed with acetate as a sole carbon source. Thus, it is necessary to further identify other important PAOs besides *Accumulibacter* in EBPR processes and to study their dynamics/interaction with GAOs and their ecophysiological traits in full-scale processes under controlled operational conditions.

4.2 IDENTITY OF PAOs

Accumulibacter. The most abundant and important Gram negative identified PAOs in most full-scale EBPR wastewater treatment plants belong to the genus *Accumulibacter* in the family *Rhodocyclaceae* of subclass 2 of the *Betaproteobacteria.* None have been grown in pure culture. *Accumulibacter* exhibit *in situ* the phenotype expected from the biochemical models for the PAO. Thus, under anaerobic conditions short-chain fatty acids are assimilated and used for synthesis of PHA with Pi release. Subsequent aerobic PHA respiration provides energy for Pi assimilation and polyphosphate synthesis. Details of their physiology are given in two reviews (Seviour *et al.,* 2003; Oehmen *et al.,* 2007). *Accumulibacter* can become highly enriched in lab-scale reactors and are also generally common in most full-scale EBPR plants, with typical relative abundances reported between 3–15% of the total biomass expressed by the EUBmix. When the functional genes of polyphosphate kinase *(ppk)* are used as phylogenetic markers at least five clades of *Accumulibacter* exist in wastewater treatment plants and natural ecosystems (Peterson *et al.,* 2008). However, 16S rRNA-targeted FISH probes do not differentiate between these clades because of the more conserved nature of this gene. Some, but not all *Accumulibacter* can denitrify, although currently the precise identity of these strains is not clear, and so it is not possible yet to recognize these with gene probes or other molecular methods.

Accumulibacter cells have a typical coccobacillus shape and are almost always observed growing in microcolonies (Figure 9.15 and 9.16, Chapter 9), although in some EBPR communities they exist predominantly as single cells.

Actinobacterial PAOs. The number of Gram positive *Actinobacteria* is often high in EBPR plants. Their functions there are largely unknown, although some of these may have a PAO phenotype. However, these do not take up short-chain fatty acids under anaerobic conditions, and their intracellular storage polymer is not PHA, but remains unidentified. Thus they do not behave according to the current biochemical models proposed for the PAO.

Two potential actinobacterial PAO morphotypes, both closely related to members of the genus *Tetrasphaera* in the family *Intrasporangiaceae* are abundant in many full-scale plants (Kong *et al.,* 2005) as a) cocci in tetrads and b) short rods. Both morphotypes *in situ* grow as relatively small cells (0.3 to 0.5 μm thick and 0.8 to 1.0 μm long) and usually form medium-sized microcolonies (Figure 9.12, Chapter 9). In some cases the cocci do not exist as the expected tetrads of *Tetrasphaera*, but form branches ('tree-shaped'), see Figure 9.13, Chapter 9. Typical rod-shaped *Tetrasphaera* are shown in Figure 9.14, Chapter 9.

Other potential PAOs. Several other bacteria grown in pure culture have been proposed to be PAOs, but none possess the expected PAO phenotype and where FISH probes have been used, were not shown to be abundant in full-scale treatment plants. These include *Acinetobacter* (Fuchs and Chen, 1975; Wagner *et al.,* 1994b), *Microlunatus phosphovorus* (Nakamura *et al.,* 1995; Santos *et al.,* 1999), *Lampropedia* spp. (Stante *et al.,* 1997), and *Tetrasphaera* spp. (Maszenan *et al.,* 2000). FISH probe-defined uncultured species of *Dechloromonas* (targeted by the probe Bet135) were proposed as putative PAOs on the basis of their phenotype. They are relatively abundant in some full-scale EBPR plants, but how important they are as PAOs is not known (Kong *et al.,* 2007).

Table 4.1. 16S rRNA oligonuclotide probes for the identification of potential PAOs.

Probe name	Target	Sequence 5′ – 3′	FA conc	Competitor oligonucleotide	Reference
PAO462	Most *Accumulibacter*	CCGTCATCTACWCAGGGTATTAAC	35	none	Crocetti *et al.*, 2000
PAO651	Most *Accumulibacter*	CCCTCTGCCAAACTCCAG	35	none	Crocetti *et al.*, 2000
PAO846	Most *Accumulibacter*	GTTAGCTACGGCACTAAAAGG	35	none	Crocetti *et al.*, 2000
PAOmix	Most *Accumulibacter*	PAO462, PAO651 and PAO846	35	none	Crocetti *et al.*, 2000
RHC439	*Rhodocyclus/Accumulibacter*	CNATTTCTTCCCCGCCGA	30	none	Hesselmann *et al.*, 1999
RHC175a	Most *Rhodocyclaceae*	TGCTCACAGAATATGCGG	30	none	Hesselmann *et al.*, 1999
PAO462b	*Rhodocyclus tenuis* group	CCGTCATCTRCWCAGGGTATTAAC	35	none	Zilles *et al.*, 2002
PAO846b	*Rhodocyclus tenuis* group	GTTAGCTACGGYACTAAAAGG	35	none	Zilles *et al.*, 2002
actino1011	*Tetrasphaera japonica*	TTGCGGGGCACCCATCTCT	30	none	Liu W. *et al.*, 2001
HGC69a	*Actinobacteria*—high G+C Gram positive bacteria	TATAGTTACCACCGCCGT	25	TATAGTTACGGCCGCCGT	Roller *et al.*, 1994
Actino221	*Actinobacteria*—potential PAOs	CGCAGGTCCATCCCAGAC	30	CGCAGGTCCATCCCATAC and CGCAGGTCCATCCCAGAG	Kong *et al.*, 2005
Actino658	*Actinobacteria*—potential PAOs	TCCGGTCTCCCCTACCAT	40	TCCGGTCTCCCCTACCAC and ATTCCAGTCTCCCCTACCAT	Kong *et al.*, 2005

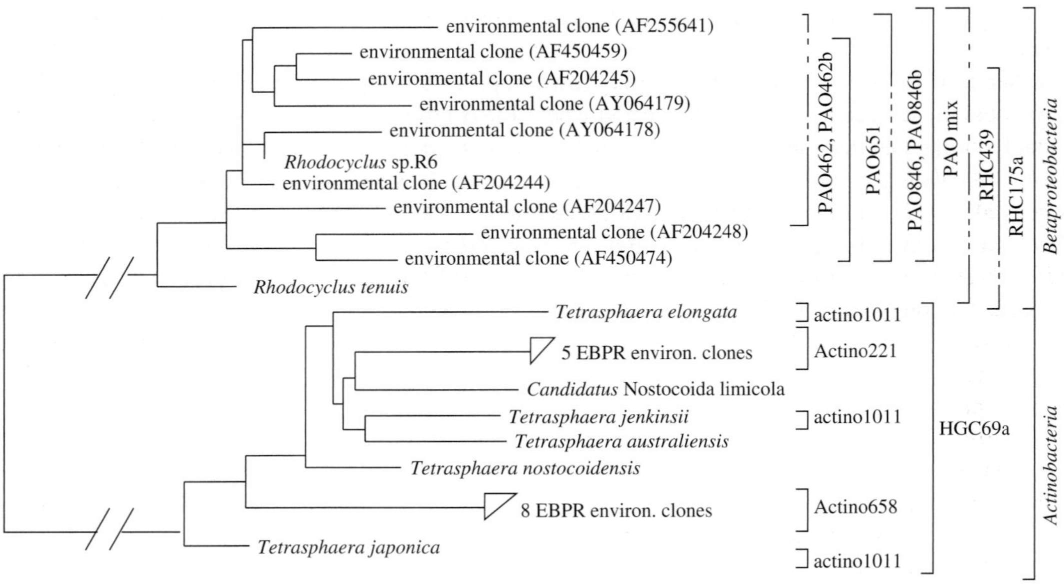

Figure 4.1. 16S rRNA-based phylogenetic tree showing main lineages of known or putative PAOs. Brackets indicate the coverage of the probes listed in Table 4.1.

4.2.1 Probes for detection of PAOs

Table 4.1 lists the oligonucleotides probes available for identifying and quantifying the PAOs or putative PAOs, and the phylogenetic tree in Figure. 4.1 illustrates the specificity and coverage of these. To quantify most *Accumulibacter* in a sample from a full-scale plant, the recommendation is to use the PAOmix probe set since it targets most currently known *Accumulibacter*-related 16S rRNA sequences. The probes of Zilles *et al.* (2002) embrace more sequences than the probes described by Crocetti *et al.* (2002) but in our experience there is no difference with full-scale plant samples compared to the results obtained when the PAOmix probe set is used.

Putative PAOs closely related to members of the genus *Tetrasphaera* can be targeted by probe Actino221 (primarily cocci in tetrads) and probe Actino658 (primarily short rods). They are also targeted by two broad-range actinobacterial probes HGC69a and actino1011. However, occasionally filamentous bacteria probably closely related to *Candidatus* Nostocoida limicola (a member of the genus *Tetrasphaera*) also fluoresce with these probes (see Chapter 5). As for most Gram positive bacteria, it may be important to apply enzymes (lysozyme 0.5 g/liter in 100 mM Tris [pH 7.5] and 5 mM EDTA) or mutanolysin (5,000 U/ml in phosphate buffer) to permeabilize cells to allow their passage to their target site, the ribosomes.

4.3 IDENTITY OF GAOs

Gammaproteobacterial GAOs. These bacteria exhibit the phenotype expected of a GAO. Namely the uptake of short-chain fatty acids and/or glucose with formation of PHA under anaerobic conditions and subsequent synthesis of glycogen but not polyphosphate under aerobic conditions. None have been grown in pure culture. Details of their physiology can be found in reviews by Seviour *et al.* (2003) and Oehmen *et al.* (2007). The most numerically abundant identified GAOs in full-scale wastewater treatment plants belong to the genus *Competibacter* within the *Gammaproteobacteria,* which are also referred to as the GB-group. The most common of these are *Candidatus* Competibacter phosphatis, often abbreviated to *Competibacter. Competibacter* can become highly enriched in lab-scale reactors but reports suggest they may also be common in full-scale EBPR plants with typical relative abundances reported up to 10% of the total cell biovolume.

More than seven subgroups of *Competibacter* are currently described, based primarily on 16S rRNA gene sequences from lab-scale reactor communities. Most of these are present or often abundant in full-scale EBPR plant communities too, but only limited information exists about possible differences in their distribution and ecology. Some *Competibacter* may denitrify but whether it is possible to distinguish these with gene probes or other molecular methods is not known.

Competibacter grow typically as coccobacilli or rods. Cells are relatively large, oval, and easily recognized under the microscope. They can form clusters of tetrads, but also appear as microcolonies (Figure 9.15, Chapter 9).

Alphaproteobacterial GAOs. Other GAOs present in some full-scale EBPR plants are members of the *Alphaproteobacteria.* Two distinct subgroups cluster 1 and 2, related to *Defluviicoccus vanus*, are distinguished based on their 16S rRNA sequences (Wong *et al.,* 2004; Meyer *et al.,* 2006). They are most commonly seen in lab-scale EBPR reactors and are generally rare in full-scale EBPR plants. Only members of cluster 2 have been seen as commonly in some EBPR plants as *Accumulibacter* and *Competibacter* (Burow *et al.,* 2007). It is likely that further phylogenetic diversity exists within this group.

The *Defluviicoccus*-related GAOs grow as cocci and rods, and typically exhibit a TFO morphology, where they form small colonies, although filamentous forms also exist (Figure 9.16, Chapter 9).

Other potential GAOs. Several cultured bacteria have also been proposed as GAOs but none are abundant in full-scale treatment plants. These include *Amaricoccus kaplicensis* (Maszenan *et al.,* 1997), also named as *Tetracoccus cechii* (Blackall *et al.,* 1997), and the original 'G' bacteria of Cech & Hartman (1993), *Quatrionicoccus* sp., *Micropruina glycogenica*, *Kineosphaera limosa* (see review by Oehmen *et al.,* 2007). Other probe-defined uncultured populations within *Betaproteobacteria* and *Gammaproteobacteria* were proposed as putative GAOs (targeted by probe Bet65 and Gam445) in some full-scale EBPR plants (Kong *et al.,* 2007) but their functional importance is unclear.

4.3.1 Probes for detection of GAOs

The oligonucleotides probes designed for identifying and quantifying these GAOs are shown in Table 4.2, and the tree in Figure 4.2 illustrates the specificity and coverage of these probes.

For most *Competibacter* in an activated sludge sample, either the general GB probe or a combination of the two lower hierarchical level probes GB_G1 probe (identical to GAOQ989) and GB_G2 can be used. Generally, the use of the GAOmix (equal amounts of GAOQ989 and GB_G2) is recommended, as this targets most known *Competibacter*-related sequences. Seven different subgroups can be targeted by GB_1-GB_7 (Kong *et al.,* 2002). However, as stated above, there is little information that supports the application of these subgroup probes, as possible differences in their distribution and ecology are not known.

Table 4.2. 16S rRNA oligonuclotide probes for the identification of potential GAOs.

Probe name	Target	Sequence 5′ – 3′	FA conc	Competitor oligonucleotide	Reference	Comment
Gam1019	Some *Gammaproteobacteria*	GGTTCCTTGCGGCACCTC	30	none	Nielsen *et al.* 1999	
Gam1278	Some *Gammaproteobacteria*	ACGAGCGGCTTTTTGGGATT	33	none	Nielsen *et al.* 1999	
GAOQ431	Some *Competibacter*	TCCCCGCCTAAAGGGCTT	35	none	Crocetti *et al.* 2002	
GAOQ989	Some *Competibacter*	TTCCCCGGATGTCAAGGC	35	none	Crocetti *et al.* 2002	
GB	Most *Competibacter*	CGATCCTCTAGCCCACT	35–70	none	Kong *et al.* 2002	
GB_G1 (GAOQ989)	Some *Competibacter*	TTCCCCGGATGTCAAGGC	35–55	TTCCCCAGATGTCAAGGC	Kong *et al.* 2002	
GB_G2	Some *Competibacter*	TTCCCCAGATGTCAAGGC	35–55	TTCCCCGGATGTCAAGGC	Kong *et al.* 2002, 2006	
GB_1 and 2	Some *Competibacter*	GGCTGACTGACCCATCC	20	none	Kong *et al.* 2002	
GB_2	Some *Competibacter*	GGCATCGCTGCCCTCGTT	35	none	Kong *et al.* 2002	
GB_3	Some *Competibacter*	CCACTCAAGTCCAGCCGT	35	none	Kong *et al.* 2002, 2006	
GB_4	Some *Competibacter*	GGCTCCTTGCGGCACCGT	35	GGCTCCTTGCGGCACCAT	Kong *et al.* 2002	
GB_5	Some *Competibacter*	CTAGGCGCCGAAGCGCCC	35	none	Kong *et al.* 2002, 2006	
GB_6 (Gam1019)	Some *Competibacter*	GGTTCCTTGCGGCACCTC	35	none	Kong *et al.* 2002	
GB_7	Some *Competibacter*	CATCTCTGGACATTCCCC	35	CATCTCTGGACGTTCCCC	Kong *et al.* 2002	
TFO_DF218	*Defluviicoccus*-related organisms (cluster 1)	GAAGCCTTTGCCCCTCAG	25–35	none	Wong *et al.* 2004	
TFO_DF618	*Defluviicoccus*-related organisms (cluster 1)	GCCTCACTTGTCTAACCG	25–35	none	Wong *et al.* 2004	
DF988	*Defluviicoccus*-related organisms (cluster 2)	GATACGACGCCCATGTCAAGGG	35	none	Meyer *et al.* 2006	Need helper H966+H1038
DF1020	*Defluviicoccus*-related organisms (cluster 2)	CCGGCCGAACCGACTCCC	35	none	Meyer *et al.* 2006	Need helper H1038

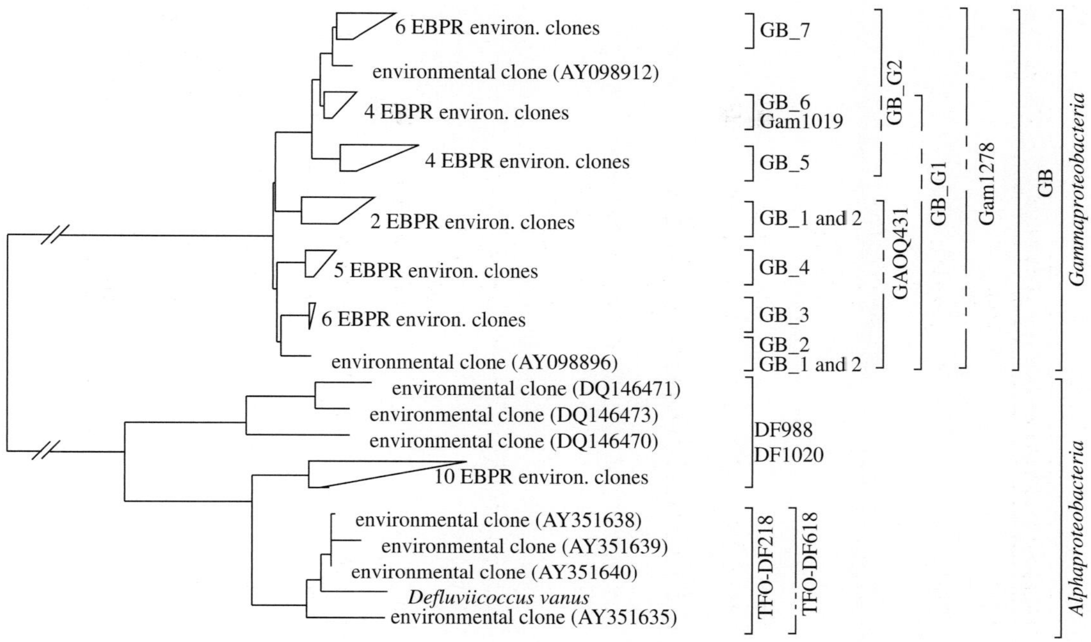

Figure 4.2. 16S rRNA-based phylogenetic tree showing main lineages of known GAOs. Brackets indicate the coverage of the probes listed in Table 4.2.

The alphaproteobacterial GAOs related to members of the genus *Defluviicoccus* can be targeted by two sets of probes. In our experience from studies of full-scale plant communities, we recommend using the DF1MIX (TFO_DF218 plus TFO_DF618) for cluster 1 and DF2MIX (DF988 plus DF1020 and helper probes H966 and H1038) for cluster 2 members.

5

Identification of filamentous bacteria by FISH

Caroline Kragelund, Elisabeth Müller, Margit Schade, Hien Thi Thu Nguyen, Hilde Lemmer, Robert Seviour, and Per Halkjær Nielsen

5.1 INTRODUCTION

Filamentous bacteria are found in all types of WWTPs, where they are often responsible for bulking (e.g. inadequate separation of biosolids and liquid effluent phases) or foaming (biosolids transported to the surface of either process tank or clarifier, typically dominated by one or two filamentous morphotypes). Filamentous bacteria should be considered as normal members of the activated sludge communities, primarily involved in degradation of organic material. Under some conditions they proliferate to such an extent that they markedly affect treatment plant performance. Their relative abundance may then exceed the usual 1–3% of the total biomass. The reasons for their excessive proliferation are many and will not be covered in this chapter, since several recent reviews and books discuss this in detail (Eikelboom, 2000; Jenkins *et al.*, 2004; Tandoi *et al.*, 2006; Nielsen *et al.*, 2009; Seviour *et al.*, 2009). In order to control the growth of these problematic bacteria, their reliable identification is necessary as the factors known to promote their growth can vary considerably. These include presence of sulphides in the influent, lack of nutrients or low oxygen concentrations in the process tank. Methods used for their attempted control are described elsewhere [for an overview on causes of sludge separation problems and control measures in several countries see (Tandoi *et al.*, 2006)]. Filamentous bacteria also inhabit biofilm systems where they are located often at their surfaces (Galvan and de Castro, 2007). Here they rarely cause problems. Bulking arising from their excessive presence may occur, but only occasionally when clarifiers are employed to

 FISH Handbook for Biological Wastewater Treatment: Identification and quantification of microorganisms in activated sludge and biofilms by FISH. Edited by Per Halkjær Nielsen, Holger Daims and Hilde Lemmer. ISBN: 9781843392316. Published by IWA Publishing, London, UK.

settle surplus biofilm sludge. Excessive growth of filamentous bacteria may also hamper settling of granules in these new reactor systems (Liu and Liu, 2006).

In 1975 Eikelboom published descriptions of a range of filamentous bacteria observed in a large number of municipal activated sludge WWTP by conventional light microscopic characterization methods (Eikelboom, 1975). He divided these into morphotypes based on morphological characteristics and staining properties, and their abundance was expressed as a numerical filament index ranging between 0 (no filamentous bacteria present) to 5 (extensive growth of filamentous bacteria), see Eikelboom (2000). Manuals containing descriptions of 26 different filament morphotypes in samples taken largely from municipal WWTPs were published subsequently and are still used routinely to identify filamentous bacteria in activated sludge (Eikelboom, 2000; Jenkins *et al.*, 2004). Furthermore, an additional 40 morphotypes have since been detected and described in industrial treatment systems (Eikelboom and Geurkink, 2000; 2002; van der Waarde *et al.*, 2002). Most of these industrial morphotypes have only rarely been found in municipal plants and only at low levels.

The presence of these morphotypes in full-scale WWTPs has been investigated and reported in several plant surveys around the world. Generally, bulking or foaming is observed in 30–50% of all plants at any given time, clearly indicating the widespread nature of these operational problems. One important outcome from these surveys is that some morphotypes appear to be associated only with some types of treatment systems defined in terms of their process design, sludge loading (sludge age) and type of influent wastewater. This probably explains why their occurrence and frequency often differ markedly between countries where different WWTP operating concepts are preferred, as demonstrated by the dominating morphotypes encountered in treatment systems in Denmark, Germany, and Australia (Table 5.1). These bacteria are divided into those seen in i) conventional plants with organic removal with or without nitrification at a sludge load of around 0.15–0.8 kg BOD/kg MLSS d (often called high F/M filaments) and ii) more advanced plants with denitrification and biological phosphorus removal at a sludge load of around 0.01–0.15 kg BOD/kg MLSS d (low F/M). Presence of filamentous bacteria and operating information from reactors with biofilms or granules is limited and not included here.

From molecular methods of analyses we now know that basing identification into morphotypes on microscopic characters is frequently inadequate for the reliable identification of many filamentous bacteria. One morphotype can embrace several phylogenetically very different organisms, as exemplified by the *Nostocoida limicola* morphotype. Studies have revealed that the *N. limicola* morphotypes affiliate with members of the class *Alphaproteobacteria* (Snaidr *et al.*, 2002), the phyla *Chloroflexi* (Schade *et al.*, 2002), *Firmicutes* (Liu *et al.*, 2000), *Planctomycetales* (Liu *et al.*, 2001b), and *Actinobacteria* (Blackall *et al.*, 2000; Liu and Seviour, 2001). This example illustrates that identification of individual filamentous organisms in activated sludge and biofilm systems often requires further molecular analyses with techniques like FISH to confirm their true identity after the manuals based on morphology have been used.

To illustrate the broad phylogenetic diversity existing among these filamentous bacteria, a comprehensive phylogenetic tree of all commonly encountered species is given in Figures 5.1a–5.1c. Most filamentous bacteria can be identified at least to phylum level by FISH (e.g. *Chloroflexi*). It is then often possible to distinguish between individual genera or species. However, more specific gene probes are still needed to completely resolve a number of morphotypes (e.g. those in Mycolata and *Chloroflexi*). In Table 5.2, we have provided an overview of the most commonly observed morphotypes and their phylogenetic affiliations when known. Furthermore, speciation of members of these phylogenetic groups where available and the corresponding morphotypes are listed in Table 5.3.

The following sections discuss each of the filamentous morphotypes described by Eikelboom (2000; 2006) (Table 5.1). They detail their abundances, provide a short description of their morphological appearances, and discuss their current phylogenetic affiliations, which are illustrated as phylogenetic trees

Table 5.1. Examples of activated sludge WWTPs and the organisms being present in Danish, German, and Australian plants. In each group the most common morphotypes are ranked.

		Danish WWTPs	German WWTPs*	Australian WWTPs
High F/M Plants with C – removal ± nitrification	**Industrial**	*Nostocoida limicola*	*Thiothrix*/type 021N	Type 021N/*Thiothrix*
		Thiothrix/type 021N	Nocardioforms/Mycolata	Mycolata
		Type 1851 and 0041/0675 without attached growth	*Nostocoida limicola* (*Alphaproteobacteria*)	*Nostocoida limicola* (*Actinobacteria* and *Alphaproteobacteria*)
	Municipal	*H. hydrossis*-like	Type 021N	Type 021N/*Thiothrix*, type 0041/0675, type 0803
		Type 1851, 1701 and 0041/0675	*S. natans*/*Leptothrix*/type 1701	*Nostocoida limicola* II
		Nocardioforms/Mycolata	Type 1863	*H. hydrossis*-like
		Thiothrix/type 021N	*H. hydrossis*-like	GALO
Low F/M Plants with N-removal ± biological P removal	**Industrial**	*Nostocoida limicola*	*Thiothrix*/type 021N	No data*
		Type 1851 and 0041/0675	Nocardioforms/Mycolata	
		Type 0803	*Nostocoida limicola* (*Alphaproteobacteria*)	
		Type 0092		
		Type 1863		
	Municipal	*H. hydrossis*-like	*Microthrix parvicella*	*M. parvicella*
		Type 1851, 1701 and 0041/0675	Type 0041/0675	Types 0092, 0914, 0041/0675
		Microthrix parvicella	Nocardioforms/Mycolata	*H. hydrossis*-like
		Type 0803	*Nostocoida limicola* II (*Actinobacteria, Chloroflexi*)	*Candidatus* Monilibacter spp.
		Type 0092	Type 1851, 0092	GALO

* Industrial WWTPs show similar filament ranking in high and low loaded plants, decisive is the carbon sources quality

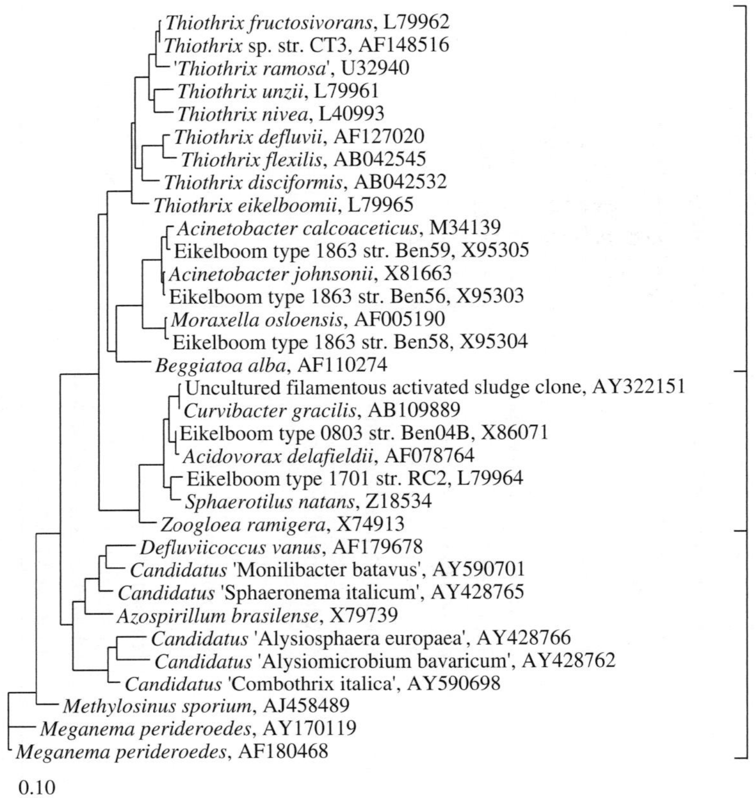

Figure 5.1a. Maximum likelihood phylogenetic tree of the 16S rRNA gene sequences of the important filamentous *Proteobacteria* and related sequences. All sequences were at least 1000 bp long. The scale bar corresponds to 0.1 substitutions per nucleotide position.

which include FISH probe coverage of each. Lastly, the currently available FISH probes for their *in situ* identification are listed and recommendations made as to which FISH probes should be used for their *in situ* identifications. However, it is important of course to check each probe specificity using available databases prior to use, since many new sequences are generated every day (see Chapter 7).

As morphotype abundance depends on plant operating parameters which can vary in different countries, we have chosen to present these filaments in alphabetic order. As many potential users of this handbook probably are familiar with the morphotype descriptions, we consider this to be the most appropriate method for its practical application.

5.2 FISH DETECTION OF FILAMENTOUS BACTERIA

General approach

In order to identify the dominant filamentous organism/s in a biofilm or activated sludge sample we propose the following procedure:

1. Observe by light microscopy (phase contrast) the abundance of filamentous bacteria and, if relevant, record the filament index (use manuals of Eikelboom, 2000; Jenkins *et al.*, 2004).

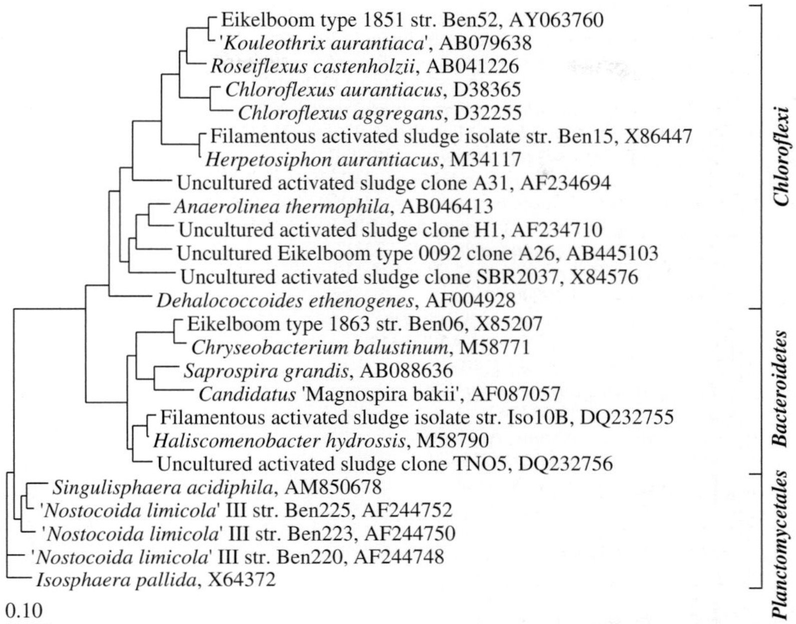

Figure 5.1b. Maximum likelihood phylogenetic tree of the 16S rRNA gene sequences of the important filamentous *Chloroflexi*, *Planctomycetales* and *Bacteroidetes* and related sequences. All sequences were at least 1100 bp long. The scale bar corresponds to 0.1 substitutions per nucleotide position.

2. Define the number of clearly distinct morphotypes (typically 1–2). Roughly 'identify' these morphotypes, e.g. *Microthrix* or *Haliscomenobacter*.
3. Go to the morphotype descriptions given in the following section (and Table 5.2) and find out which species each morphotype is thought to cover.
4. If it only covers one distinct morphotype (e.g. *Microthrix parvicella*), use the recommended probe(s) for FISH detection (i.e. MPAmix). Further details about other probes can be found within the description section.
5. If more than one species falls within that single morphotype (e.g. *Haliscomenobacter*): Select a number of broad gene probes that covers each potential species (e.g. one for phylum *Bacteroidetes* and one for phylum *Chloroflexi*), see Table 5.4. When the correct phylum is found, more narrow probes can then be used (e.g. HHY-654 for *H. hydrossis* in *Bacteroidetes* or EU25-1238 for Type 1851 in *Chloroflexi*; see respective Tables and Figures).
6. Include always the EUBmix probe tagged with a different fluorochrome to ensure that the filament is detectable by the FISH protocol. If no fluorescence signal is obtained with the EUBmix the filament may be dead or the probes do not penetrate the cell wall (e.g. for Mycolata) and a pretreatment is necessary. Some *Chloroflexi* filaments do not hybridize with the EUBmix probes because they lack the probes' target sites (Kragelund *et al.*, 2007a). **It is crucial to make sure that the morphology and the FISH signal from the selected probes coincide**. Use the images presented in this book and those given in other manuals for comparison. Always include a negative control when examining a new sample (nonsense probe) to detect any autofluorescence from the sample (this is not uncommon!).

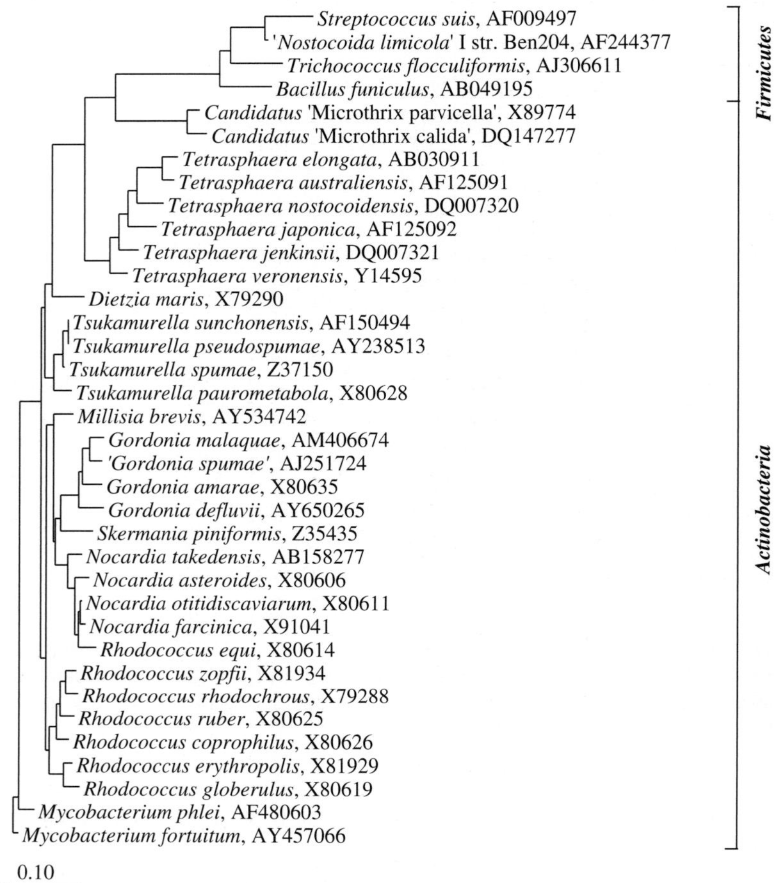

Figure 5.1c. Maximum likelihood phylogenetic tree of the 16S rRNA gene sequences of the important filamentous Gram positive bacteria and related sequences. All sequences were at least 1300 bp long. The scale bar corresponds to 0.1 substitutions per nucleotide position.

5.3 FILAMENTOUS BACTERIA DETECTION AND IDENTIFICATION

This section provides advice on how best to select the FISH probes to detect and identify filamentous bacteria found in activated sludge, foam, and biofilms. It includes actual data on their abundance in municipal and industrial WWTPs and biofilm reactors, a brief morphological description, their phylogenetic affiliation, and recommendations for FISH probes best suited for this purpose.

For *in situ* monitoring purposes a semiquantitative filament count might be assessed by determining a class index according to Eikelboom (2000). This is done for both conventional microscopy and FISH in order to achieve comparable data. Class indices from 0 indicating no filamentous growth to 5 indicating extensive growth of filamentous organisms with class index steps differing by a factor of 10 turned out to be helpful for practitioners' monitoring needs. In case of FISH overall biomass concentrations should be checked by DAPI or a universal bacteria probe such as EUBmix to be similar to those examined in wet mount.

Table 5.2. Overview of the most common morphotypes and their possible phylogenetic identity.

Morphotype	Possible phylogenetic identity
Beggiatoa	*Beggiatoa* spp. (class *Gammaproteobacteria*)
Haliscomenobacter hydrossis	*Haliscomenobacter hydrossis* (*Bacteroidetes*) Other species in the phylum *Bacteroidetes* Species within the phylum *Chloroflexi*
Leucothrix mucor	*Leucothrix* sp. (class *Gammaproteobacteria*)
Microthrix parvicella	*Candidatus* M. parvicella (*Actinobacteria*) *Candidatus* M. calida (*Actinobacteria*) Unknown species within the phylum *Chloroflexi*
Nostocoida limicola	Several species of *Alphaproteobacteria* *Candidatus* N. limicola (*Actinobacteria*) *Thiothrix* (type 021N) species (class *Gammaproteobacteria*) Species within the phylum *Firmicutes* Species within the phylum *Chloroflexi* Species within the phylum *Planctomycetes*
Nocardioform actinomycetes/Mycolata	*Gordonia* species and *G. amarae* *Skermania* species and *S. piniformis* *Rhodococcus* species, *Rhodococcus erythropolis* *Dietzia* species Unknown species
Sphaerotilus natans/ Leptothrix discophora	*Sphaerotilus* spp. (class *Betaproteobacteria*) *Leptothrix* spp. (class *Betaproteobacteria*)
Thiothrix	*Thiothrix* (class *Gammaproteobacteria*) Species within the phylum Candidate division TM7 without attached growth Species within the phylum *Chloroflexi* without attached growth
Type 0041/0675	Species within the phylum *Chloroflexi* *Curvibacter*-related (class *Betaproteobacteria*) Candidate division TM7
Type 0092	Unknown *Chloroflexi* (class *Chloroflexi*) Perhaps species within the phylum *Bacteroidetes*
Type 021N	*Thiothrix* species (class *Gammaproteobacteria*) *Meganema perideroedes* (class *Alphaproteobacteria*) *Candidatus* Nostocoida limicola (*Actinobacteria*) Species within the phylum *Chloroflexi* Species within the phylum *Firmicutes*
Type 0803/0914	Unknown Perhaps species in the class *Betaproteobacteria* or phylum *Chloroflexi*
Type 1701	*Curvibacter*-related (class *Betaproteobacteria*) *Sphaerotilus*-related (class *Betaproteobacteria*) Unknown
Type 1851	*Kouleothrix aurantiaca* (phylum *Chloroflexi*) Species within the phylum *Chloroflexi* *Curvibacter*-related (*Betaproteobacteria*)
Type 1863	*Acinetobacter* spp. (class *Gammaproteobacteria*) *Chryseobacterium* spp. (*Bacteroidetes*)

Table 5.3. Overview of the phylogenetic affiliation, species name if possible and corresponding morphotype.

Phylogenetic identity Phylum/class	**Genus/Species name**	**Morphotype**
Proteobacteria/ Alphaproteobacteria	*Candidatus* Alysiosphaera europaea	*N. limicola*
	Candidatus Monilibacter batavus	
	Candidatus Alysiomicrobium bavaricum	
	Candidatus Combothrix italica	
	Meganema perideroedes	Type 021N
Proteobacteria/ Betaproteobacteria	Genus *Curvibacter*	Type 1701
	Genus *Curvibacter*	Type 0041/0675
	Sphaerotilus natans	*Sphaerotilus natans*
	Leptothrix discophora	*Leptothrix*
Proteobacteria/ Gammaproteobacteria	*Thiothrix nivea, T. unzii, T. fructosivorans, T. defluvii*	*Thiothrix* species
	T. eikelboomii, T. disciformis, T. flexilis	Type 021N species
	Acinetobacter spp.	Type 1863
	Leucothrix mucor	*Leucothrix mucor*
	Beggiatoa spp.	*Beggiatoa*
Bacteroidetes	*Haliscomenobacter hydrossis*	*H. hydrossis*
	Other *H. hydrossis*-like species	*H. hydrossis*-like
	Unknown species	Type 0092
	Chryseobacterium spp.	Type 1863
Chloroflexi	*Kouleothrix aurantiaca*	Type 1851
	Unknown species	*Nostocoida limicola* II
	Other *Chloroflexi*, (species not identified)	Thin *H. hydrossis*-like (÷epiphytic growth)
	Other *Chloroflexi*, (species not identified)	Type 1851 (+epiphytic growth)
	Other *Chloroflexi*, (species not identified)	Type 0041/0675 (±epiphytic growth)
Actinobacteria	*Candidatus* M. parvicella	*M. parvicella*
	Candidatus M. calida	Thin *M. parvicella*
	Candidatus Nostocoida limicola	*Nostocoida limicola* II
	Genus *Gordonia* and *G. amarae*	Mycolata (GALO)
	Genus *Skermania* and *S. piniformis*	Mycolata (PTLO)
	Genus *Rhodococcus*	Mycolata (GALO)
	Genus *Dietzia*	Mycolata (NOC-like)
	Unknown Mycolata	Unknown Mycolata
Candidate division TM7	TM7_genera_incertae_sedis	Type 0041/0675 (±epiphytic growth)

Table 5.3. *Continued*

Phylogenetic identity → morphotype		
Phylogenetic identity Phylum/class	**Genus/Species name**	**Morphotype**
Firmicutes	*Trichococcus* spp. *Streptococcus* spp.	*Nostocoida limicola* I *Streptococcus*
Planctomycetales	*Isosphaera* spp.	*Isosphaera*, see text
Unknown	Unknown Unknown	Type 0803/0914 Type 0961 and others see text

5.3.1 *Beggiatoa* morphotype

Abundance, morphology and affiliation

Abundance of *Beggiatoa* morphotype reported in plants in Australia, the Netherlands and the US (summarized in Jenkins *et al*., 2004) is moderate. No bulking or foaming problems have been implied from its presence. They are often present in low numbers in biofilms (Nielsen PH and Nielsen JL, unpublished results). *Beggiatoa* spp. serve as a strong indicator for a presence of reduced sulfur compounds, as it may be mixotrophic, using these as energy sources.

The *Beggiatoa* morphotype is characterized by being mobile by gliding and exhibiting straight to bent filaments with a cell diameter of 1.5–2.5 μm. *In vivo* sulfur granules are often visible. *Beggiatoa* stain Gram negative and Neisser negative.

The genus *Beggiatoa* is affiliated in the family *Thiotrichaceae* (class *Gammaproteobacteria*).

Recommending the best probes

Three *Beggiatoa* strains have been isolated from activated sludge (Williams and Unz, 1985) but no phylogenetic analyses were performed on them. Only FISH probes for marine species exist (Mussmann *et al*., 2003; Macalady *et al*., 2006), and so the broad class probe GAM42a for *Gammaproteobacteria* might indicate the presence of *Beggiatoa* in activated sludge, see Table 5.4. However, *Beggiatoa* is identifiable routinely by phase contrast microscopy only. The probes for marine species have never resulted in a FISH positive signal for this morphotype in activated sludge samples (Nielsen JL and Kragelund C, unpublished results).

5.3.2 *Haliscomenobacter hydrossis* morphotype

Abundance, morphology and affiliation

The abundance of *H. hydrossis* is reported in many surveys based on morphology (summarized in Jenkins *et al*., 2004), usually without ever being present in large amounts. Surveys using FISH probes have confirmed that *H. hydrossis* is almost always present in minor amounts in both municipal and industrial plants, independent of plant configuration (Kragelund *et al*., 2008).

The *H. hydrossis* morphotype exists as short straight (needle-like appearance) or longer slightly bent filaments, usually protruding from the flocs. Filament length is variable, and trichome diameter is approx. 0.4 μm. Cells are surrounded by a sheath. They stain Gram negative and Neisser negative.

These distinctively thin needle-shaped bacteria usually affiliate with members of *H. hydrossis* in the genus *Haliscomenobacter* (family *Saprospiraceae*) within the phylum *Bacteroidetes*. However, other

Table 5.4. General probes covering the different phyla and classes or broader groups. Can be used together with more specific probes or, if no specific probes exist, to reveal the overall affiliation of an otherwise unidentified filamentous organism.

Probe name	Target	Morphotype	Sequence (5′-3′)	FA (%)	Comments	Reference
EUBmix (EUB-338 + EUB338-II# + EUB338-III)	Most *Bacteria*, *Planctomycetales and Verrucomicrobiales*		GCT GCC TCC CGT AGG AGT + GCA GCC ACC CGT AGG TGT + GCT GCC ACC CGT AGG TGT	0–60	Both filaments and single cells	Amann *et al.* 1990; Daims *et al.* 1999
ALF968	class *Alphaproteobacteria*	*N. limicola*	GGT AAG GTT CTG CGC GTT	35	Both filaments and single cells	Neef, 1997
GAM42a	class *Gammaproteobacteria*	*Beggiatoa, Thiothrix, Leucothrix*	GCC TTC CCA CAT CGT TT	35	Use together with BET42a	Manz *et al.* 1992
BET42a	class *Betaproteobacteria*	*Leptothrix, S. natans* are	GCC TTC CCA CTT CGT TT	35	Use together with GAM42a	Manz *et al.* 1992
CFB563	Most *Flavobacteria*	*H. hydrossis*	GGA CCC TTT AAA CCC AAT	20	Both filaments and single cells	Weller *et al.* 2000
CF319a+b	Most *Flavobacteria*	*H. hydrossis*	TGG TCC GTG TCT CAG TAC+ TGG TCC GTA TCT CAG TAC	35	Both filaments and single cells	Manz *et al.* 1996
CFB286	Most members of the genus *Tannerella* and the genus *Prevotella* of the class *Bacteroidetes*	*H. hydrossis*	TCC TCT CAG AAC CCC TAC	50	Both filaments and single cells	Weller *et al.* 2000
CFB719	Most members of the class *Bacteroidetes*	*H. hydrossis*	AGC TGC CTT CGC AAT CGG	30	Both filaments and single cells	Weller *et al.* 2000
LGC354B	*Firmicutes* (Gram positive bacteria with low G+C content)	*Streptococcus, N. limicola* I	CGG AAG ATT CCC TAC TGC	35	Together with LGC 354A and C. Both filaments and single cells	Meier *et al.* 1999
LGC354A	*Firmicutes*	*Streptococcus, N. limicola* I	TGG AAG ATT CCC TAC TGC	35	Together with LGC 354B and C. Both filaments and single cells	Meier *et al.* 1999
LGC354C	*Firmicutes*	*Streptococcus, N. limicola* I	CCG AAG ATT CCC TAC TGC	35	Together with LGC 354A and B. Both filaments and single cells	Meier *et al.* 1999
Pla46	*Planctomycetales*	*N. limicola*	GAC TTG CAT GCC TAA TCC	30	Both filaments and single cells	Neef *et al.* 1998
HGC69a+ comp	*Actinobacteria*	GALO/PTLO, *N. limicola*	TAT AGT TAC CAC CGC CGT+ TAT AGT TAC GGC CGC CGT	25	Both filaments and single cells	Roller *et al.* 1994; 1995

HGC1156	*Actinobacteria*	*Microthrix parvicella*	CGAGTTGACCCCGGCAGT	20	Both filaments and single cells*	Erhart *et al.* 1997
Myc657	*Mycobacterium* sub-division Mycolata (mycolic acid-containing actinomycetes)	GALO/PTLO	AGT CTC CCC TGY AGT A	30	Both filaments and single cells*	Davenport *et al.* 2000
MNP1	Mycolata (actinomycetes, *Corynebacterineae*)	GALO	TTA GAC CCA GTT TCC CAG GCT	50	Both filaments and single cells**	Schuppler *et al.* 1998
CMN119	Suborder CMN excluding *Dietzia* spp. and *Tsukamurella* spp.	GALO/PTLO	GGCAGATCACCCACGTGT	30	Both filaments and single cells***	Erhart, 1997
CFX mix (GNSB941 +CFX1223)	Phylum *Chloroflexi*	Type 0041/0675, 1851, *H. hydrossis*	AAACCACACGCTCCGCT+ CCATTGTAGCGTGTGTGTMG	35	Also other morphotypes can be observed, often type 1851 and sometimes also type 1701, and *H. hydrossis*-like filaments	Bjornsson *et al.* 2002 Gich *et al.* 2001
TM7905	Nearly the entire candidate division TM7	Type 0041/0675	CCGTCAATTCCTTTATGTT TTA	20	Also other morphotypes are detected: thick filaments, rods, and coccoid cells	Hugenholtz *et al.* 2001

Formamide 0–35%
* A. use pretreatment with achromopeptidase and lysozyme, no extended hybridization time needed
** B. use lysozyme, acid, lipase, proteinase K and extended hybridization time 15–18h
*** C. Lysozyme and extended hybridization time 15–18h

filaments with the same morphology are often detected, either belonging to other members of the *Bacteroidetes* (Kragelund *et al.*, 2008) or to those in the phylum *Chloroflexi* (Kragelund *et al.*, 2007a). Other filamentous bacteria can demonstrate a similar appearance but they have been identified solely on morphology and their phylogenetic affiliations are currently unresolved.

Recommending the best probes

Table 5.5 shows those probes for identification and quantification of the *H. hydrossis* morphotype, and the tree presented in Figure 5.2 illustrates their specificity and coverage.

Table 5.5. ***Haliscomenobacter hydrossis*** **morphotype:** rRNA-targeted oligonucleotide probes with corresponding formamide concentration, target, morphotype, and comments on recommendability.

Probe name	Target	Morphotype	Sequence (5′-3′)	FA (%)	Comments	Reference
SAP-309	Most members of *Saprospiraceae*	*H. hydrossis*	TCT CAG TAC CCG TGT GGG	25	Both filaments and single cells	Schauer and Hahn, 2005
HHY-654	*H. hydrossis* and Isolate 10B	*H. hydrossis*	GCC TAC CTC AAC YTG ATT	35		Kragelund *et al.* 2008
HHY	*H. hydrossis*	*H. hydrossis*	GCC TAC CTC AAC CTG ATT	20–25		Wagner *et al.* 1994a

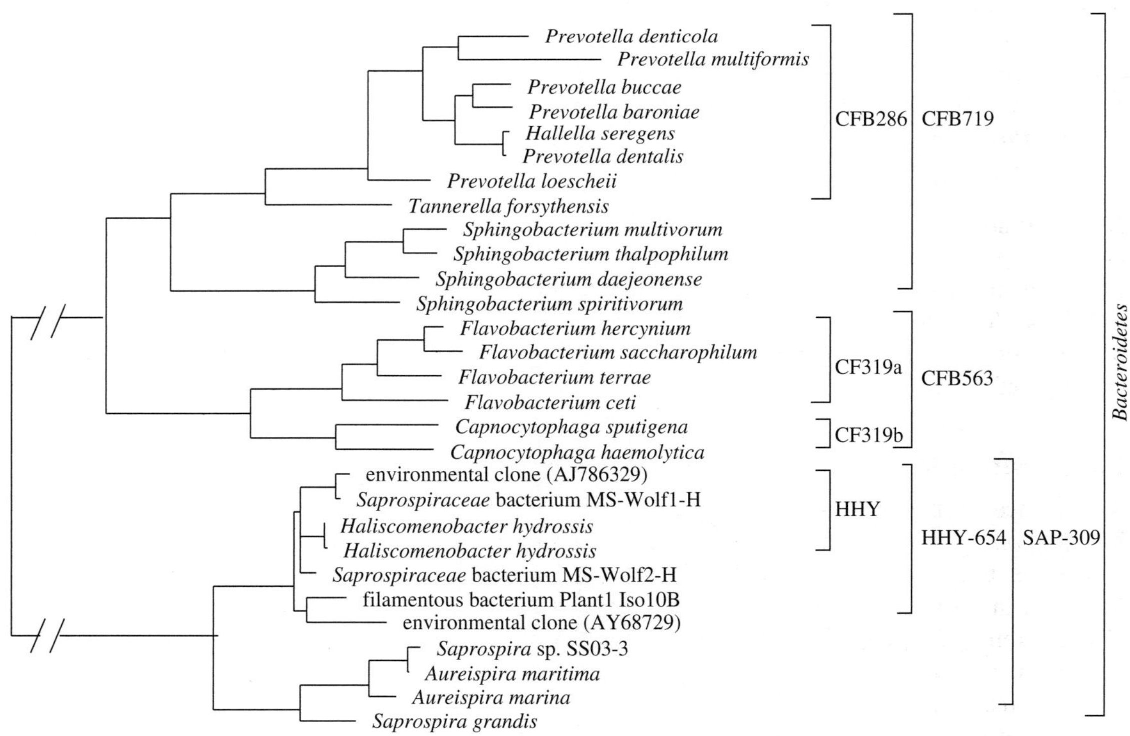

Figure 5.2. 16S rRNA-based phylogenetic tree showing main lineages of known *H. hydrossis* bacteria. Brackets indicate the coverage of the probes listed in Table 5.5. The exact branching order of the *H. hydrossis* lineages cannot yet be determined due to a limited phylogenetic resolution of the 16S rRNA marker gene. Therefore, unclear tree topology is depicted as multifurcation.

The most commonly used probe is HHY, targeting the species *H. hydrossis* which was designed against a pure culture isolate (Wagner *et al.*, 1994a). A broader probe HHY-654 is now available which also targets an additional *H. hydrossis* isolate (Kragelund *et al.*, 2008). Substantially more *H. hydrossis* respond to this probe in full-scale plants. Other probes have been published to detect *Bacteroidetes* in different habitats but these are very broad and do not target individual species (e.g. CFB286, CFB563 and CFB719, Weller *et al.*, 2000). Even though these were designed originally to target freshwater or marine *Bacteroidetes* (CFB286) they also target filaments in activated sludge. Other probes detect members of the phylum *Bacteroidetes* (CF319 a+b, Manz *et al.*, 1996), targeting both single cells and filamentous bacteria. The SAP-309 probe (Schauer and Hahn, 2005) targets members of the family *Saprospiraceae* which contain both single cells and filamentous bacteria.

Our recommendation for detection of the *H. hydrossis* morphotype in the *Bacteroidetes* in activated sludge is probe HHY-654. To detect most *H. hydrossis* and *H. hydrossis*–like bacteria, the broader targeting probes CFB719 and SAP-309 can be used in combination. Morphological appearance of probe-defined *H. hydrossis* morphotype can be seen in FISH image Figure 9.17, Chapter 9.

5.3.3 *Leucothrix mucor* morphotype

Abundance, morphology and affiliation

The abundance of *Leucothrix mucor* has been assessed in surveys based only on morphology (Jenkins *et al.*, 2004). It has also been detected by FISH in industrial samples (van der Waarde *et al.*, 2002). It is observed only occasionally in WWTPs where it plays no decisive role in bulking or foaming incidences (van der Waarde *et al.*, 2002; Jenkins *et al.*, 2004).

The *Leucothrix mucor* morphotype is characterized by bent to twisted filaments, a highly variable cell diameter and a variable cell morphology of discoid, round, to rod-shaped cells. It never has intracellular S-granules and stains Gram negative and Neisser negative.

Leucothrix mucor is member of the genus *Leucothrix*, closely related to genus *Thiothrix* in the family *Thiotrichaceae* (class *Gammaproteobacteria*).

Recommending the best probes

Only one probe exists for *L. mucor* (LMU) (Wagner *et al.*, 1994a) designed based on a pure culture isolate. It can also be targeted by the class specific probe for *Gammaproteobacteria* (GAM42a, see Table 5.4). The LMU probe sequence is included in Table 5.6, which contains FISH probes available for less commonly observed filament morphotypes. The tree in Figure 5.3 illustrates the specificity and coverage of these probes.

5.3.4 *Microthrix parvicella* morphotype

Abundance, morphology and affiliation

The *Microthrix parvicella* morphotype is an important bacterium responsible for bulking and foaming incidences throughout the world (Eikelboom, 2000; Jenkins *et al.*, 2004). Two species have been identified in activated sludge. One, *Candidatus* M. parvicella, predominates in WWTPs with biological N and P removal treating primarily domestic wastewater (Blackall *et al.*, 1996). The other, *Candidatus* M. calida, resembles this morphotype but its trichome is slightly thinner and it can grow at higher temperatures in pure culture. It has only been found so far in WWTPs treating mainly industrial waste, both with and without biological N and P removal (Levantesi *et al.*, 2006).

Microthrix parvicella morphotype filaments are bent to curled with a cell diameter of 0.5 – 0.6 μm (*M. calida*: cell diameter between 0.3 and 0.6 μm). They stain Gram positive and Neisser positive with

Table 5.6. Morphotype *Leucothrix mucor, Sphaerotilus natans, Leptothrix discophora*, *Streptococcus* and 1863 morphotype: rRNA-targeted oligonucleotide probes with corresponding formamide concentration, target, morphotype, and comments on recommendability. Exemplified for different morphotypes which are not very commonly observed in activated sludge.

Probe name	Target	Morphotype	Sequence (5′-3′)	FA (%)	Comments	Reference
LMU	*Leucothrix mucor* (*Gammaproteobacteria*)	*Leucothrix*	CCC CTC TCC CAA ACT CTA	35		Wagner *et al.* 1994a
SNA	*Sphaerotilus natans* (*Betaproteobacteria*)	*S. natans*	CAT CCC CCT CTA CCG TAC	45		Wagner *et al.* 1994a
CTE	Competitor for SNA		TTC CAT CCC CCT CTG CCG	20		Schleifer *et al.* 1992
LDI	*Leptothrix discophora* (*Betaproteobacteria*)	*Leptothrix*	CTC TGC CGC ACT CCA GCT	35		Wagner *et al.* 1994a
Strept*	*Streptococcaceae* (*Firmicutes*)	*Streptococcus*	CAC TCT CCC CTT CTG CAC	40	pretreatment C	Trebesius *et al.* 2000
ACA23a	*Acinetobacter spp.* (*Gammaproteobacteria*)	Type 1863	ATC CTC TCC CAT ACT CTA	35		Wagner *et al.* 1994b

* This probe is denoted Str in probeBase. C. Lysozyme and extended hybridization time 15–18h.

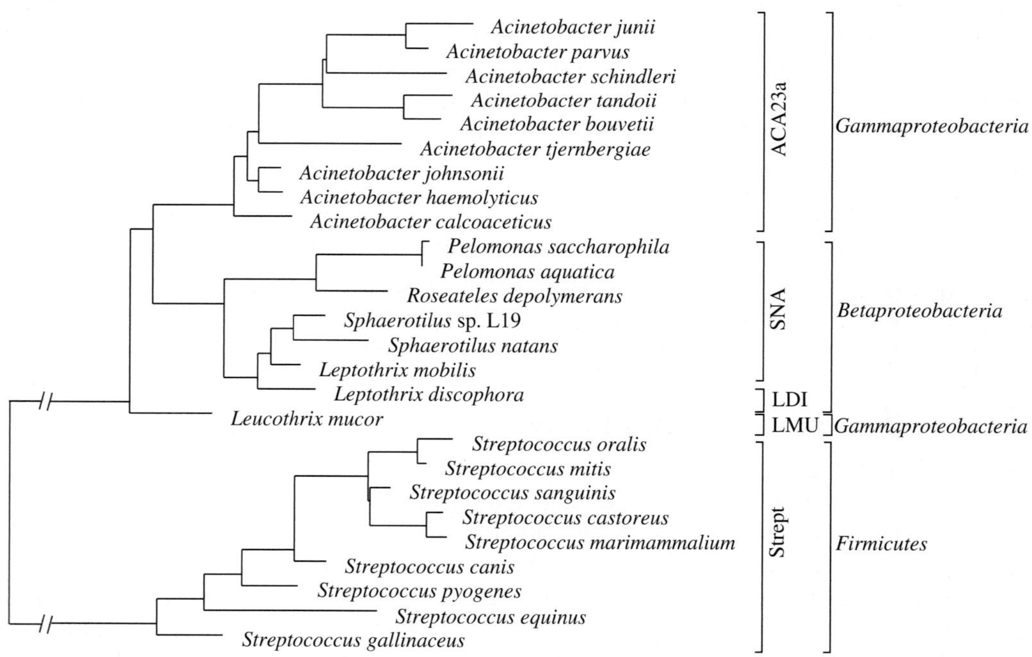

Figure 5.3. 16S rRNA-based phylogenetic tree showing main lineages of different morphotypes which are not very commonly observed in activated sludge. Brackets indicate the coverage of the probes listed in Table 5.6. The exact branching order of the bacterial lineages cannot yet be determined due to a limited phylogenetic resolution of the 16S rRNA marker gene. Therefore, unclear tree topology is depicted as multifurcation.

polyP granules, which may be absent, for example, after prolonged starvation periods or after treatment with polyaluminum products (Eikelboom, 2006). *M. parvicella* morphotype filaments can also contain lipid/PHB granules (Rossetti *et al.*, 2005).

Candidatus M. parvicella and *Candidatus* M. calida both affiliate with the *Actinobacteria* (unclassified *Actinobacteria*) (Blackall *et al.*, 1996).

Recommending the best probes

Table 5.7 shows the probes currently available for the morphotype *M. parvicella*, and the tree in Figure 5.4 illustrates their specificity and coverage.

To detect *M. parvicella*, cell permeabilization by pretreatment with enzymes is required to obtain a satisfactory FISH signal. Many protocols have been published for permeabilizing *M. parvicella* and other Gram positive bacteria (Beimfohr *et al.*, 1993; Erhart *et al.*, 1997; Carr *et al.*, 2005; Kragelund *et al.*, 2007b). That which seems to promote the strongest FISH signal is combining achromopeptidase and lysozyme (Kragelund *et al.*, 2007b), which was developed originally for CARD-FISH (Sekar *et al.*, 2003). An extended hybridization time of up to 15–18 hours might be needed to eliminate irregular fluorescence signal strength along the filament (Müller *et al.*, 2007).

The phylum probe for *Actinobacteria* (HGC69a) designed to target their 23S rRNA (Roller *et al.*, 1994) is not suitable for *Candidatus* M. parvicella, since at least two mismatches exist with it (Bradford *et al.*, 1998). Instead the probe HGC1156 (Erhart, 1997) targeting the *Actinobacteria* can be used for *Candidatus* M. parvicella (Müller *et al.*, 2007) and *Candidatus* M. calida. Several probes have been published that target the *M. parvicella* morphotype. Four probes MPA645, MPA223, MPA60, and MPA650 were developed for *in situ* detection of *Candidatus* M. parvicella (Erhart *et al.*, 1997). The latter probe requires two competitor probes. Another two probes Mpa-all-1410 and Mpa-T1-1260 were developed which target *Candidatus* M. calida (Levantesi *et al.*, 2006). However, Mpa-all-1410 has a weak central mismatch to *Candidatus* M. parvicella, but despite this mismatch, the probe clearly hybridized to *Candidatus* M. parvicella isolates and *M. parvicella* present in sludge samples.

We recommend the MPAmix (mixture of MPA645, MPA223 and MPA60) be used for *in situ* detection of the *Microthrix* morphotype including both *Candidatus* M. parvicella and *Candidatus* M. calida, together with the HGC1156 probe targeting *Actinobacteria*. To distinguish between similar

Table 5.7. *Microthrix* morphotype: rRNA-targeted oligonucleotide probes with corresponding formamide concentration, target, morphotype and comments on recommendability.

Probe name	Target	Morphotype	Sequence (5′-3′)	FA (%)	Comments	Reference
MPAmix (MPA60 + MPA223 + MPA645)	*Candidatus* M. parvicella + *Candidatus* M. calida	*Microthrix parvicella*	GGA TGG CCG CGT TCG ACT + GCC GCG AGA CCC TCC TAG + CCG GAC TCT AGT CAG AGC	20	Use a probe mix and pretreatment A or B*	Erhart *et al.*, 1997
Mpa-all-1410	Almost all sequences belonging to *Candidatus* M. calida	Thin and normal *M. parvicella*	GGT GTT GTC GAC TTT CGG CG	35	pretreatment A or B*	Levantesi *et al.*, 2006
Mpa-T1-1260	All sequences belonging to *Candidatus* M. calida	Thin *M. parvicella*	TTC GCA TGA CCT CAC GGT TT	25	pretreatment A or B*	Levantesi *et al.*, 2006

A. use pretreatment with achromopeptidase and lysozyme, no extended hybridization time needed
B. use lysozyme, acid, lipase, proteinase K and extended hybridization time 15–18h

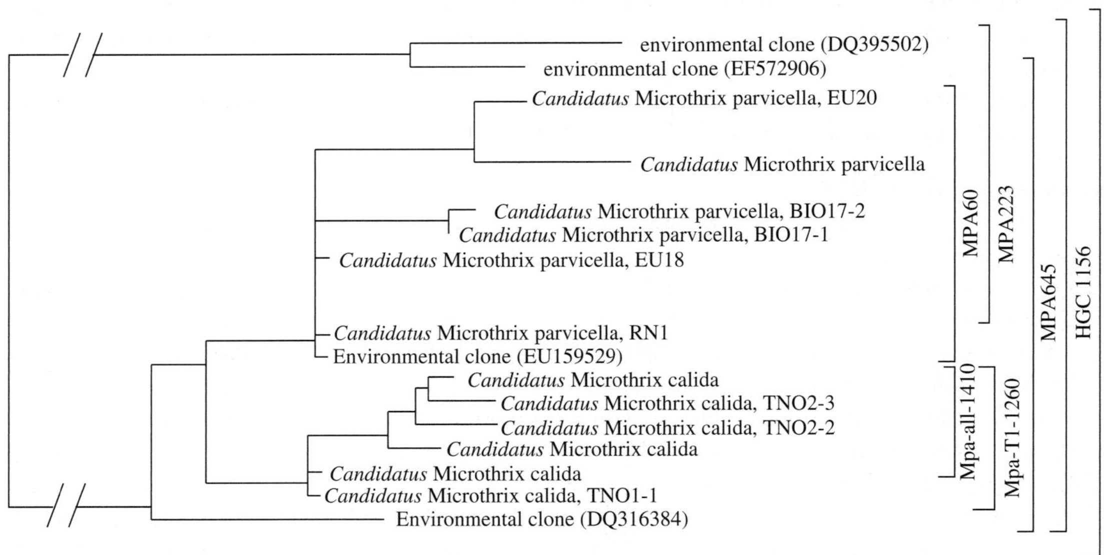

Figure 5.4. 16S rRNA-based phylogenetic tree showing main lineages of *Microthrix* morphotypes in the *Actinobacteria*. Brackets indicate the coverage of the probes listed in Table 5.7. The exact branching order of the *Actinobacteria* lineages cannot yet be determined due to a limited phylogenetic resolution of the 16S rRNA marker gene. Therefore, unclear tree topology is depicted as multifurcation.

looking *Candidatus* M. parvicella and *Candidatus* M. calida in a sludge sample, MPAmix should be applied first in order to detect both. Thereon Mpa-T1-1260 should be used to visualize *Candidatus* M. calida filaments only. FISH images of *M. parvicella* and *M. calida* morphotypes can be seen Figures 9.18 and 9.19 in Chapter 9.

5.3.5 *Nostocoida limicola* morphotypes

Abundance, morphology and affiliation

The abundance of the different *N. limicola* morphotypes (I–III) has been assessed by morphologically based surveys and these are associated with bulking and foaming (Eikelboom, 2000; Jenkins *et al.*, 2004). Several surveys using FISH probes have been published; Probe NLIMII175 detected this morphotype frequently in WWTPs of Australia, New Zealand, France (Liu and Seviour 2001) and in a higher frequency than did the NLII65 probe in German WWTPs (Müller *et al.*, 2007). The *N. limicola* affiliating with the *Alphaproteobacteria* seem to play a major role in the bulking events in plants with a high load of industrial waste (Levantesi *et al.*, 2004). In municipal WWTPs in Germany only a few have been recorded (Müller *et al.*, 2007), while in Australian EBPR plants they are more abundant (Seviour, unpublished). The *Chloroflexi* affiliated *N. limicola* morphotype has been observed in several German scums in municipal plants (Müller *et al.*, 2007).

The *N. limicola* morphotype is distinguished by its long, coiled and tangled filaments. Three different morphotypes (I, II, and III) have been described based on minor differences primarily in their cell diameter (Eikelboom, 2000). Individual variable opaque coccoid cells are barely resolvable by light microscopy in thin filaments of *N. limicola* type I. Cells are coccoid to discoid in the thicker type II filaments. Flattened discoid cells are found in *N. limicola* type III. Diameter varies from ca. 0.8 to 2 μm. *Nostocoida limicola* filaments are Gram variable and often but not always Neisser positive, and show a light to dark grey violet cell wall.

It is difficult (or impossible) to relate each of the morphotypes to phylogenetically distinct probe-defined species. They have been identified as members of several phylogenetically distantly related species. Some affiliate with the *Alphaproteobacteria*, others with members of the *Chloroflexi, Firmicutes, Planctomycetes,* and *Actinobacteria.*

Recommending the best probes for *Nostocoida limicola* II

Table 5.8 shows probes that can be used for *N. limicola* morphotypes, and the tree in Figure 5.5 illustrates their specificity and coverage.

***Nostocoida limicola* morphotype II** is commonly seen in municipal WWTPs. Many belong to *Candidatus* N. limicola affiliating with the genus *Tetrasphaera* in the family *Intrasporangiaceae* in the phylum *Actinobacteria.* This filamentous species is usually not observed in large quantities in bulking incidences but often in low numbers especially in municipal WWTPs. Importantly, in German plants the NLIMII175 probe has been observed to also target other morphotypes including bent sickle-shaped rods, and long thin filaments developing two different morphotypes with long rectangular cells and spherical to square-shaped cells (Schade M., unpublished). A typical *N. limicola* morphotype can be seen in Fig 9.20 in Chapter 9.

The *N. limicola* morphotype II can also be a member of the phylum *Chloroflexi* (class *Thermomicrobia*, Hugenholtz and Stackebrandt, 2004) and is targeted by probe AHW183 (designed based on 4 cultured isolates, Schade *et al.*, 2002). A typical *N. limicola* morphotype can be seen in Figure 9.21 in Chapter 9.

In industrial WWTPs the most commonly observed *N. limicola* morphotypes II belong to the *Alphaproteobacteria,* and these account for many incidents of bulking (Levantesi *et al.*, 2004). The most commonly observed filaments *Candidatus* Monilibacter batavus, *Candidatus* Alysiomicrobium bavaricum, *Candidatus* Sphaeronema italicum and *Candidatus* Alysiosphaera europaea can be detected with the general probe (ALF968) for the *Alphaproteobacteria*, and more specifically with probes MC2-649 and DF198, PPx3-1428, Sita-649 and Noli-644, respectively. These probes are not very specific, for example, MC2-649 (Nittami, unpublished results), PPx3-1428 and Sita-649 (Kragelund *et al.*, 2006), but they can still be used for preliminary screening. For some of these filaments more specific probes are available such as PPx1002 for *Candidatus* Alysiomicrobium bavaricum, and Nost993 in combination with a helper probe for *Candidatus* Sphaeronema italicum (Kragelund *et al.*, 2006). Probe DF988 designed for the GAO *Defluviicoccus* (see Chapter 4) also targets *Candidatus* Monilibacter batavus (Seviour R., unpublished results). Other alphaproteobacterial species with this morphology have been identified in activated sludge: *Meganema perideroedes* can be detected with probes Meg983 and Meg1028 (Thomsen *et al.*, 2006a) and *Candidatus* Combothrix italica is detectable with probe Combo1013 (Levantesi *et al.*, 2004). However, these filaments are not commonly observed in either industrial or municipal WWTPs.

For identification of these alphaproteobacterial *Nostocoida limicola* morphotype II we recommend the following probes listed in Table 5.8. The ALF968 probe detects all known filaments belonging to the *Alphaproteobacteria* (see Table 5.4). For further species differentiation the following probes are recommended: DEF198, PPx3-1428, Sita-649 and Noli-644 which target the most commonly observed species predominantly in industrial sludges (Table 5.8). However, bear in mind that more specific probes are available for some of the alphaproteobacterial species and please consult the listed references for details. For the actinobacterial species, probe NLIMII175 is recommended, but be aware that other

Table 5.8. *Nostocoida limicola* morphotypes: rRNA-targeted oligonucleotide probes with corresponding formamide concentration, target, morphotype, and comments.

Probe name	Target	Morphotype	Sequence (5′-3′)	FA (%)	Comments	Reference
NLIMII175	*Candidatus* Nostocoida limicola II strains except strain Ben 70 (phylum *Actinobacteria*)	*Nostocoida limicola* II	GGC TCC GTC TCG TAT CCG	40		Liu and Seviour, 2001
NLII65	Nostocoida limicola II (phylum *Actinobacteria*)	*Nostocoida limicola II*	CAA GCT CCT CGT CAC CGT T	20		Bradford, 1997
AHW183	*N. limicola*-like isolate, nearest relative *Sphaerobacter thermophilus* (*Chloroflexi* phylum)	*Nostocoida limicola* II	CCGACACTACCCACTCGT	35		Schade *et al.* 2002
PPx3-1428	*Candidatus* Alysiomicrobium bavaricum (Alisphaera ppx3) (Class *Alphaproteobacteria*)	*Nostocoida limicola* II	TGG CCC ACC GGC TTC GGG	50	Optional hybridization overnight	Snaidr *et al.* 2002
Noli-644	*Candidatus* Alysiosphaera europaea (Class *Alphaproteobacteria*)	*Nostocoida limicola* II	TCC GGT CTC CAG CCA CA	35	Optional hybridization overnight	Snaidr *et al.* 2002
Sita-649+ Comp Sita-649	*Candidatus* Sphaeronema italicum + competitor (Class *Alphaproteobacteria*)	*Nostocoida limicola* II	CCW CTC CCG GAC YCC AGC + CCT CTC CCG GTC TCC AGC C	50		Levantesi *et al.* 2004
DF198	*Candidatus Monilibacter batavus*-related organisms (Class *Alphaproteobacteria*)	*Nostocoida limicola* II	ATCCCAGGGCAACATAGT CT	35		Nittami *et al.* unpublished
Combo1031	*Candidatus* Combothrix italica (Class *Alphaproteobacteria*)	*Nostocoida limicola* II	CAC CTG CAG TGG CCT CCC GA	35		Levantesi *et al.* 2004
Meg1028 + Meg983	All published sequences of *Meganema perideoedes* (Class *Alphaproteobacteria*)	Type 021N	CTG TCA CCG AGT CCC TTG C + CGG GAT GTC AAA AGG TGG	35		Thomsen *et al.* 2006
NLIMI91	*Nostocoida. limicola* I isolates, *Trichococcus* spp. (*Firmicutes*)	*Nostocoida limicola* I	CGCCACTATCTTCTCCGT (published sequence) CGCCACTATCTTCTCAGT (correct sequence)	20	Lysozyme	Liu and Seviour, 2001
NLIMIII301	*N. limicola* III isolate, *Isosphaera pallida* (*Planctomycetales*)	*Nostocoida limicola* III	CCCAGTGTGCCGGGCCAC	20	Lysozyme	Liu and Seviour, 2001

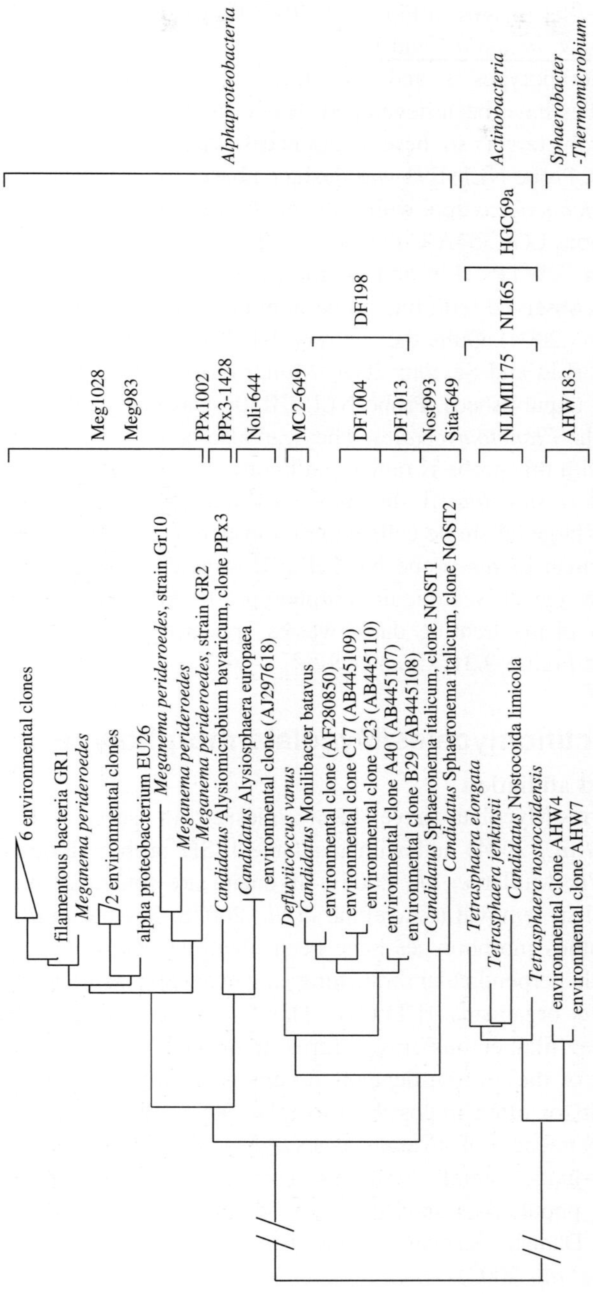

Figure 5.5. 16S rRNA-based phylogenetic tree showing main lineages of *Nostocoida limicola* morphotypes. Brackets indicate the coverage of the probes listed in Table 5.8. The exact branching order of the *Nostocoida limicola* morphotypes lineages cannot yet be determined due to a limited phylogenetic resolution of the 16S rRNA marker gene. Therefore, unclear tree topology is depicted as multifurcation.

morphotypes have been observed with this probe. Also the phylum probe HGC1156 targets the morphotype (Table 5.4). Lastly, the *Chloroflexi*-related N. limicola II is targeted by probe AHW183 which is also targeted by CFX mix (Table 5.4). Morphological appearance of *Nostocoida limicola* II morphotypes can be seen in FISH images in Figures 9.20–9.30, Chapter 9.

Other probes for morphotypes *N. limicola* I and III

For *Nostocoida limicola* morphotypes I and III affiliating to members of the *Firmicutes* and *Planctomycetales* several probes have been developed. However, there are some discrepancies between morphotype definitions and probe targets so these do not necessarily target *N. limicola* morphotypes I and III as described by Eikelboom. Probe NLIMI91 was designed based on sequences from three *N. limicola*-like isolates, affiliated with *Trichococcus* spp. within the phylum *Firmicutes* (Liu *et al.*, 2000). It can also be targeted by the phylum probes LGC354A-C (Table 5.4). This probe was used to detect *N. limicola* I in Australian, Italian and French WWTPs (Liu and Seviour, 2001). However, no detection of the typical *N. limicola* I morphotype was observed with this probe in either municipal or industrial European plants (Eikelboom, 2006; Müller *et al.*, 2007). Quite the contrary, NLIMI91 detected short filaments and coccoid cells in Australia and Europe (Liu and Seviour 2001, Müller *et al.*, 2007) and long thin filaments with rectangular cells (Schade M., unpublished). Probe NLIMIII301 was designed based on five *N. limicola* III-like isolates belonging to the *Planctomycetales*. They can also be targeted by the phylum probe Pla46 (Table 5.4). The experience with this probe is rather conflicting: In contrast to the opaque cells described by Eikelboom for morphotype *Nostocoida* III, they possessed transparent cocci in slightly bent filaments (Liu *et al.*, 2001b). The probe targeted single cells within transparent *Isosphaera* filaments in some cases (Müller *et al.*, 2007). In German plants probe NLIMIII301-targeted cells were never present in high abundance as compared to the typical N. limicola morphotype III (Müller *et al.*, 2007), but were highly abundant in some Australian plants treating dairy wastes appearing as long chains of coccoid cells (Seviour R., unpublished), see Figure 9.31, Chapter 9.

5.3.6 Nocardioform actinomycetes/Mycolata morphotype

Abundance, morphology and affiliation

Abundances of Nocardioforms/Mycolata have been assessed in surveys based on morphology (Eikelboom, 2000; Jenkins *et al.*, 2004) and with available FISH probes (Eales *et al.*, 2005; Müller, 2006; Kragelund *et al.*, 2007b; Müller *et al.*, 2007). Mycolata are found in municipal and industrial WWTPs with or without nutrient removal (Kragelund *et al.*, 2007b; Seviour *et al.*, 2008).

Two quite distinct filamentous morphotypes have been identified. *Gordonia amarae*-like organisms (GALOs) have filaments with perpendicular branching angles, while acute-angle branched filaments distinguish the pine tree-like organisms (PTLOs). This difference is used for their microscopic 'identification'. In addition, non-filamentous growth forms may occur, and here are referred to as NOC-like organisms. Development of these often depends on environmental conditions or organism growth phase. Therefore it is difficult (or often impossible) to relate any of these to phylogenetically distinct probe-defined populations. Abundances of several species of Mycolata have been assessed using available FISH probes in foam and sludge samples. Studies have examined the abundances of *Gordonia, Skermania,* and *Rhodococcus* populations and NOC-like organisms in plant samples from all over the world (e.g. Australian, Czech, Danish, German, and Swedish plants) (Eales *et al.*, 2005; Kragelund *et al.*, 2007b; Müller, 2006; Müller *et al.*, 2007).

Their cell diameters vary from 0.6–2.0 μm, depending on the morphotype present. Usually they are Gram positive, but Gram variable morphotypes have also been observed especially in industrial plants or those operating at a very high sludge age. Neisser/DAPI positive polyP granules are occasionally seen.

The Mycolata morphotypes formerly known as 'nocardioforms' all contain mycolic acid in their cell walls. They are classified as members of the suborder *Corynebacterineae* within the *Actinobacteria* (Stackebrandt *et al.*, 1997). This group encompasses members of several genera including *Corynebacterium, Dietzia, Gordonia, Millisia, Mycobacterium, Nocardia, Rhodococcus, Skermania, Tsukamurella,* and *Williamsia,* based primarily on chemical, molecular, and morphological markers (Chun *et al.*, 1997; Butler *et al.*, 2005; Soddell *et al.*, 2006). *Nocardia amarae* has been reclassified as *Gordonia amarae* (Goodfellow *et al.*, 1994; Klatte *et al.*, 1994; Ruimy *et al.*, 1994). *Nocardia pinensis* has also been reclassified into a new genus *Skermania* as *Skermania piniformis* (Chun *et al.*, 1997). Phylogenetic and phenotypic studies of Mycolata in WWTPs reveal a high diversity among members of this group and *Dietzia, Gordonia, Mycobacterium, Rhodococcus, Tsukamurella*, and *S. piniformis* are often involved in foaming events (Goodfellow *et al.*, 1998; Schuppler *et al.*, 1998; Soddell *et al.*, 1998; Seong *et al.*, 1999; Stainsby *et al.*, 2002; Wagner and Cloete, 2002; Nam *et al.*, 2003; Eales *et al.*, 2005; Müller, 2006; Kragelund *et al.*, 2007b).

Recommending the best probes

In general FISH protocols for detecting Mycolata incorporate a pretreatment permeabilization step to facilitate probe transport into the cells, as described above for the *Microthrix* morphotype. A lysozyme pretreatment at least is recommended. For some *Rhodococcus* spp. and *Skermania piniformis* an acid, lipase, proteinase K treatment might be necessary (Carr *et al.*, 2005, Müller, 2006). The protocol that seems to promote the strongest FISH signal for detecting *Gordonia* and *Skermania* species is combining achromopeptidase and lysozyme (Kragelund *et al.*, 2007b). In addition extending hybridization times to 15–18 h is recommended, as outlined earlier for *Microthrix*.

Table 5.9 shows the FISH probes useful for Mycolata morphotypes at group, genus, and species level, and the tree in Figure 5.6 illustrates their specificity and coverage. Firstly probes for detecting the entire Mycolata group are described, and then those targeting individual genera and species are discussed.

Mycolata group-targeted probes. No probe has been designed that targets the entire Mycolata group. Probe HGC1156 targets most known Mycolata but also many other members of the phylum *Actinobacteria*. However, branched filaments (GALOs and PTLOs) can to some extent be detected with the probe if no intermediate morphotypes are observed as reported in several papers (Kragelund *et al.*, 2007b). But NOC-like organisms cannot be distinguished from other non-mycolic acid-containing organisms. Therefore, additional group- and genus-targeted probes are necessary for further identification. This step is essential since individual Mycolata members can differ markedly in their physiological features, indicating that individual control strategies are required for each.

Two broad probes (Myc657 and MNP1) target most known members of the family *Corynebacteriaceae* where most of the genera of mycolic acid containing branched filaments are grouped (Myc657; Davenport *et al.*, 2000; MNP1; Schuppler *et al.*, 1998), see Table 5.4. Probe Myc657 detects more Mycolata than MNP1, but interestingly the latter probe detects NOC-like organisms better than the branched filamentous bacteria. Probe CMN119 (Erhart *et al.*, 1997) detects a subgroup of the *Corynebacterineae* excluding *Dietzia* spp. and *Tsukamurella* spp. and this probe has identified GALOs which could not be detected with the other broad probes (Müller, 2006). These broad probes are grouped in Table 5.4 along with other group or phylum probes.

Gordonia (GALO)-specific probes. The probe Gor596 is suitable for detecting all known members of the genus *Gordonia* (de los Reyes *et al.*, 1997). Other probes have been designed to target *Gordonia* spp. in sludge and foam (Gam192, *G. amarae* (de los Reyes *et al.*, 1997), GLP2, different *Gordonia* sequences

Table 5.9. Nocardioform actinomycetes/Mycolata morphotype: rRNA-targeted oligonucleotide probes with corresponding formamide concentration, target, morphotype, and comments on recommendability.

Probe name	Target	Morpho-type	Sequence (5′-3′)	FA (%)	Comments	Reference
Gor596	*Gordonia* family	GALO	TGC AGA ATT TCA CAG ACG ACG C	20	Pretreatment A	de los Reyes *et al.* 1997
G.am205	*Gordonia amarae*	GALO	CAT CCC TGA CCG CAA AAG C	30	Pretreatment A	de los Reyes *et al.* 1998
Spin1449	*Skermania piniformis*	PTLO	CCGCTCCCTCCCACAAAG	35	Pretreatment A	Eales *et al.* 2006
R.ery619	*Rhodococcus erythropolis*	GALO	CCTGCAAGCCAGCAGTTG	20	Pretreatment C	Müller, 2006
RHOa429	*Rhodococcus* spp. Cluster A scum clone *Rhodococcus* spp.	GALO	CGGAGCTGAAAGGAGTTT	20	Pretreatment C	Müller, 2006
RHOb183	*Rhodococcus* sp. Cluster B scum clone	GALO	ACCACGAAACATGCATCC	20	Pretreatment C	Müller, 2006
DIE993	*Dietzia* spp.	NOC-like org.	CCGTCGTCCTGTATATGT	20	Pretreatment C	Müller, 2006

A. use pretreatment with achromopeptidase and lysozyme, no extended hybridization time needed
C. Lysozyme and extended hybridization time 15–18h

(Schuppler *et al.*, 1998), and G.am205, *G. amarae* (de los Reyes *et al.*, 1998). Hierarchical probe overlap is observed with Myc657 and MNP1.

Skermania piniformis (PTLO)-specific probes. Two probes for detecting *Skermania piniformis* have been published: Spin1449 (Eales *et al.*, 2006), and NPI452 (Bradford, 1997). PTLOs are commonly observed using probe Spin1449 in foaming European WWTPs. Hierarchical probe overlap is observed with Myc657.

Rhodococcus (GALO)-specific probes. Members of the genus *Rhodococcus* are a diverse group of organisms. Rainey *et al.* (1995) showed that *Rhodococcus* species fall into six distinct phylogenetic groups and currently contain some members whose taxonomic status is uncertain. Therefore it is not possible to design a probe targeting all members of this genus. Nevertheless, two probes Rco1 and Rco2 are described which detect most *Rhodococcus* spp. (Davenport *et al.*, 1998). However, these two probes showed no positive FISH results with pure cultures of *Rhodococcus fascians, R. rhodnii, R. globerulus*, and *R. rhodochrous* (Müller E., unpublished). They were optimized under different FISH conditions than used here (e.g. different hybridization formamide concentrations and temperature) and are thus not recommended.

Three probes are available for different *Rhodococcus* spp. The probe R.ery619 targets *R. erythropolis* (Müller, 2006) (*Rhodococcus* group IV according to Rainey *et al.*, 1995). Hierarchical probe overlap occurs with Myc657 and MNP1 (Müller, 2006). The other two probes, i.e. RHOa429 and RHOb183, were designed to detect two *Rhodococcus* clusters not affiliated to any of Rainey's *Rhodococcus* groups or any known *Rhodococcus* species. Probe RHOa429 targets *Rhodococcus* spp. cluster A, and RHOb183 targets *Rhodococcus* sp. cluster B. Sequence analyses of *Rhodococcus* cluster A and B members show one and two mismatches to the MNP1 and Myc657 probes, respectively. They respond only to the group-specific probe CMN119 (Erhart, 1997; Müller, 2006).

Dietzia (NOC-like organisms)-specific probes. Two probes are available against *Dietzia* spp. showing a NOC-like appearance in scum. Probe DLP (Schuppler *et al.*, 1998) identifies members of the genus

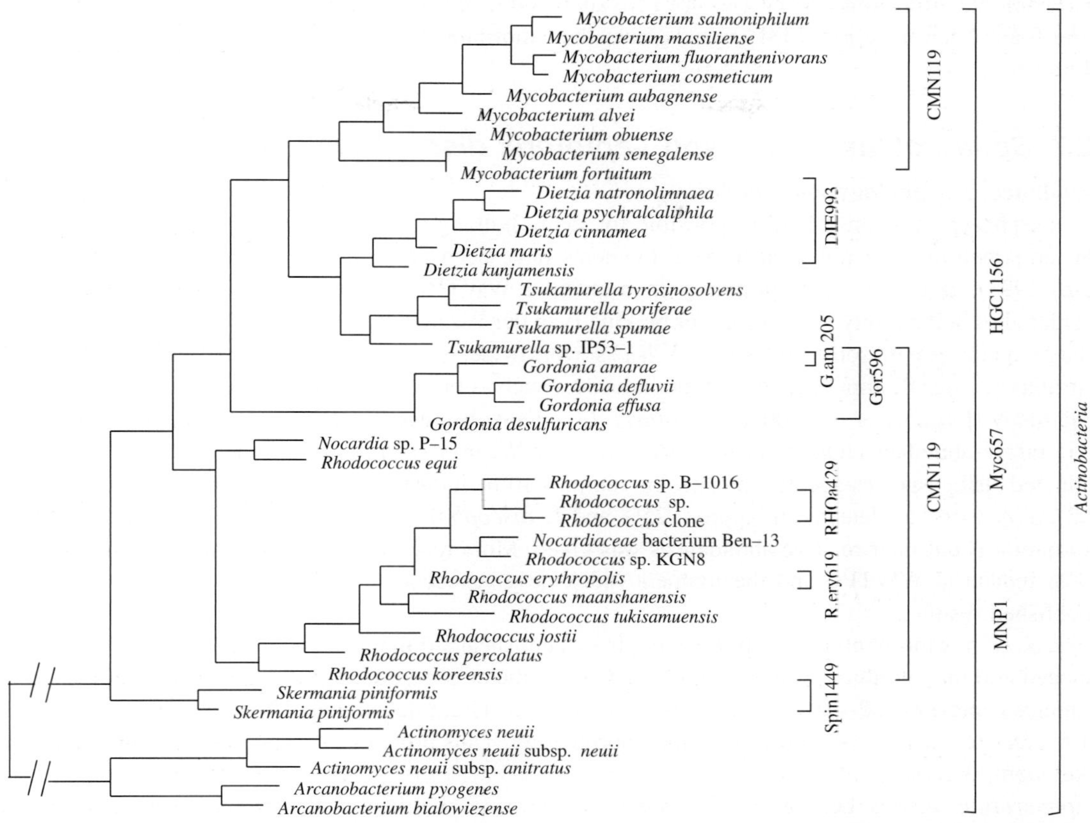

Figure 5.6. 16S rRNA-based phylogenetic tree showing main lineages of Nocardioform actinomycetes/ Mycolata morphotype. Brackets indicate the coverage of the probes listed in Table 5.9. The exact branching order of the Nocardioform actinomycetes/Mycolata lineages cannot yet be determined due to a limited phylogenetic resolution of the 16S rRNA marker gene. Therefore, unclear tree topology is depicted as multifurcation.

Dietzia (excluding *D. maris*), which form short irregular rods (Schuppler *et al.*, 1998). Another probe DIE993 targets *D. maris*, *D. psychralcaliphila*, *D. daqingensis*, *D. natronolimnaea*, uncultured foam bacteria (Wagner and Cloete, 2002), and scum clones (Müller, 2006), all of which are members of the genus *Dietzia*. The morphology of the NOC-like organisms was bent single rods either in clusters or forming short non-branched filaments. Hierarchical probe overlap is observed with Myc657 and also MNP1 (Müller, 2006).

Recommending the best probes. We recommend probe Myc657 as the general probe targeting most Mycolata. For detecting *Rhodococcus* spp. in sludge or foam it might be advantageous to apply CMN119 as a general probe also because some *Rhodococcus* species are not targeted by the Myc657 probe. At genus level the following probes Gor596 (*Gordonia* spp.), Spin1449 (*Skermania piniformis*), RHOa429, RHOb183, R.ery619 (three different *Rhodococcus* spp.) and DIE993 (*Dietzia* spp.) differentiate between some common species within the broad Mycolata group. However, these more specific probes do not cover the entire Mycolata group and so it is still not possible to identify all GALOs and PTLOs by FISH.

For reliable identification it is best to use activated sludge samples and not foam samples, as organisms in the foam often have low FISH signals from their moribundity. See FISH images, Figures 9.32–9.38 in Chapter 9.

5.3.7 *Sphaerotilus natans* and *Leptothrix discophora* morphotype

Abundance, morphology and affiliation

This morphotype was once held responsible for many bulking incidences in plants operating with high F/M ratios before other potential bulking filaments have been described (e.g. Eikelboom 1975, Jenkins *et al.*, 1993). It is still seen, primarily in municipal WWTPs without nutrient removal, but is now considered as being rarely responsible for bulking (van der Waarde *et al.*, 2002). However, it can be found in large quantities in some industrial WWTPs worldwide, such as the pulp and paper industries in Australia (Seviour R, unpublished) or food industries such as breweries, dairies or starch processing plants in Germany (Lemmer *et al.*, 2000). *Leptothrix discophora* has been identified in activated sludge plants but is rarely abundant (Jenkins *et al.*, 1993; van der Waarde *et al.*, 2002). When a FISH survey was conducted using species-specific probes on samples from industrial WWTPs (van der Waarde *et al.*, 2002), *S. natans* was detected in approx. 25% and *L. discophora* in less than 1% of these plants, but no information about their relative abundances was given. More recently *S. natans* was present in only 13% of 126 industrial WWTPs, and the filament index never exceeded 1.5 (Kragelund C and Nielsen PH, unpublished results).

The *S. natans* morphotype comprises straight or bent sheathed filaments which are distinctively falsely branched and may produce swarming gonidia. Cell diameter is approx. 1.0–1.3 μm. Cells are rectangular (filaments center) or rod-shaped (filaments ends). It stains Gram negative and Neisser negative. *S. natans* and *L. discophora* share an almost identical morphology. Thus probes are needed for their differentiation (Eikelboom, 2006).

Sphaerotilus natans belongs to the genus *Sphaerotilus* within the *Betaproteobacteria*. *Leptothrix discophora* is also a member of the *Betaproteobacteria* and is located within the genus *Leptothrix*.

Recommending the best probes

Probes SNA and LDI (Wagner *et al.*, 1994a) for detecting the *S. natans* and *L. discophora* morphotypes, respectively, are described in Table 5.6, which contains sequences for less commonly seen filament morphotypes, while the tree in Figure 5.3 illustrates their specificity and coverage. These FISH probes were both developed with pure cultures of *S. natans* and *L. discophora*. Probe SNA for *S. natans* requires the CTE competitor probe, since Wagner *et al.* (1994a) report that *Sphaerotilus* spp., *S. natans,* type 1701, *L. discophora,* and *L. cholodnii* all hybridized with it. Moreover, type 1701 and *Leptothrix* spp. show a close phylogenetic affiliation with *Sphaerotilus* spp. In German municipal WWTPs the probe SNA even with CTE still visualized single rod-shaped and coccoid cells in addition to *Sphaerotilus* filaments (Schade M., unpublished). Both *S. natans* and *L. discophora* can be hybridized by the class-specific probe for *Betaproteobacteria* (Table 5.4). See FISH images, Fig 9.39, Chapter 9.

5.3.8 *Streptococcus* morphotype

Abundance, morphology and affiliation

Filamentous streptococci can be quite abundant in municipal WWTPs with EBPR, constituting 1–4% of the biovolume (Kong *et al.*, 2008). The filaments of *Streptococcus* are bowed and twisted, round cells with a diameter of approx. 0.7 μm. Septa and constrictions are clearly visible. They stain Gram positive and Neisser negative. Filamentous streptococci affiliate in the genus *Streptococcus* in the family *Streptococcaceae* (class *Bacilli,* phylum *Firmicutes*). This morphotype may also cover other *Firmicutes*, such as *Trichococcus* spp. (Liu *et al.*, 2000), which can also be assessed as morphotype *Nostocida limicola* I, see section above.

Probes for detection of streptococci

Table 5.6 shows that probe Strept (Trebesius *et al.*, 2000) can be used to identify filamentous streptococci (see also Chapter 6). The probe targets 44% of all streptococci, and can be used together with the broader probe LGC354C (Meier *et al.* 1999). The morphologies of the Strept-probed cells embrace both small cocci and filamentous organisms. The phylogenetic tree in Figure 5.3 shows the less commonly seen filament morphotypes and the probes for their detection along with their specificity and coverage.

5.3.9 *Thiothrix* and Type 021N morphotypes

Abundance, morphology and affiliation

These morphotypes occur in municipal and industrial WWTPs with and without nutrient removal (Eikelboom, 2000; Jenkins *et al.*, 2004) and in biofilm reactors (Nielsen PH. and Nielsen JL., unpublished results). FISH surveys have applied several probes targeting these morphoptypes and they were seen frequently in industrial samples where they were found in 38% of samples (detected using probes TNI and 21N) (van der Waarde *et al.*, 2002). More specific probes targeting subpopulations of morphotype 021N were used in a survey, and two of the three 021N morphotypes were frequently seen in Japanese and European WWTPs (Kanagawa *et al.*, 2000).

The *Thiothrix* morphotypes contain six microscopically recognizable different forms, whose trichomes can be separated on their diameter, which may vary from 0.6–2.5 μm with rectangular cells in the filament center and rod-shaped cells at their ends. Often sulfur granules accumulate *in situ* (Unz and Head, 2005). They generally display a positive S^0 result without or after incubation with a reduced sulfur source. They are Gram and Neisser negative.

The 021N filament morphotype has a cell diameter of 1.0–2.0 μm with discoid or almost square cells in the centre and rod-shaped cells at the filament ends. Occasionally small sulfur granules are visible *in situ* or after incubating with a reduced sulfur source. Type 021N stains Gram and Neisser negative.

Several morphologically indistinguishable filaments have been identified: at least one belongs to the *Alphaproteobacteria* (*Meganema perideroedes*) and possibly others (Howarth *et al.*, 1999; Thomsen *et al.*, 2006a) and some affiliate to the *Chloroflexi* being not separable microscopically from *T. eikelboomii* (Type 021N) (Kragelund *et al.*, 2007a). As the S^0 test is not always useful in identifying each, FISH probes are needed to differentiate between them.

Currently, seven species of *Thiothrix* in the *Gammaproteobacteria* have been isolated and described (Cole *et al.*, 2005; 2008; Unz and Head, 2005) and furthermore *T. nivea* (Teske *et al.*, 1996) and *T. ramosa* (Polz *et al.*, 1996) have been extensively described. The Eikelboom 021N morphotype has now been shown to represent three separate species within the genus *Thiothrix* (*T. eikelboomii, T. disciformis* and *T. flexilis*) (Howarth *et al.*, 1999; Kanagawa *et al.*, 2000; Unz and Head, 2005).

Recommending the best probes

Table 5.10 shows the oligonucleotide probes for identifying and quantifying the *Thiothrix*/Type 021N group, and the tree in Figure 5.7 illustrates their specificity and coverage.

Table 5.10. *Thiothrix*/021N morphotype rRNA-targeted oligonucleotide probes with corresponding formamide concentration, target, morphotype, and comments on recommendability.

Probe name	Target	Morphotype	Sequence (5′-3′)	FA (%)	Comments	Reference
G123T + G123T-C competitor	*Thiothrix*/021N group (*Gammaproteobacteria*)	Type 021N & *Thiothrix* spp.	CCT TCC GAT CTC TAT GCA + CCT TCC GAT CTC TAC GCA	40		Kanagawa *et al.* 2000
TNI	*Thiothrix nivea*	*Thiothrix*	CTC CTC TCC CAC ATT CTA	45		Wagner *et al.* 1994a
21N	Type 021N	Type 021N	TCC CTC TCC CAA ATT CTA	35		Wagner *et al.* 1994a
G1B	021N group 1 (*T. disciformis*)	Type 021N	TGT GTT CGA GTT CCT TGC	30		Kanagawa *et al.* 2000
G2M	021N group 2 (*T. eikelboomii*)	Type 021N	GCA CCA CCG ACC CCT TAG	35		Kanagawa *et al.* 2000
G3M	021N group 3 (*T. flexilis*)	Type 021N	CTC AGG GAT TCC TGC CAT	30		Kanagawa *et al.* 2000
Meg1028 + Meg983	*Meganema perideroedes* (*Alphaproteobacteria*)	Type 021N	CTG TCA CCG AGT CCC TTG C + CGG GAT GTC AAA AGG TGG	35		Thomsen *et al.* 2006a

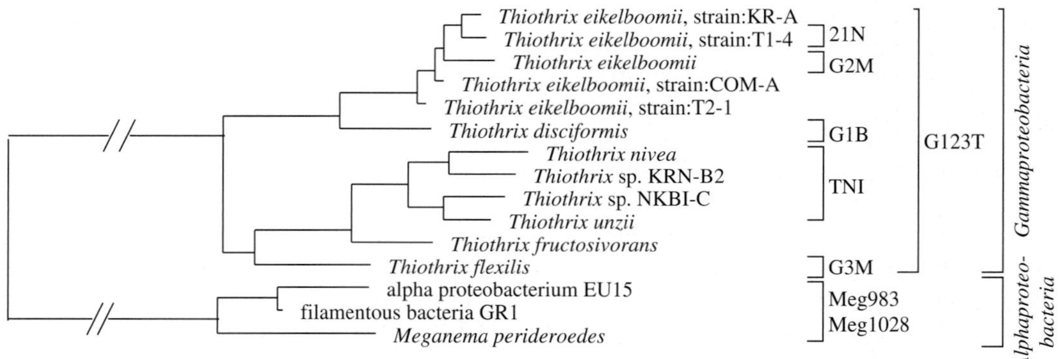

Figure 5.7. 16S rRNA-based phylogenetic tree showing main lineages of *Thiothrix*/O21N. Brackets indicate the coverage of the probes listed in Table 5.10. The exact branching order of the *Thiothrix*/O21N lineages cannot yet be determined due to a limited phylogenetic resolution of the 16S rRNA marker gene. Therefore, unclear tree topology is depicted as multifurcation.

The probe G123T with its competitor G123T-C (Kanagawa *et al.*, 2000) targets the entire *Thiothrix*/021N group. Probes 21N and TNI (Wagner *et al.*, 1994) target subpopulations of type 021N and *Thiothrix*, respectively. The latter also hybridizes with some rod-shaped and coccoid cells, presumably the gonidia from *Thiothrix*. Probe 21N occasionally targets much thinner filaments, which are readily distinguished from type 021N (Schade M., unpublished). Probes based on pure culture isolates allow the three type 021N species to be differentiated: G1B targets group 1 (*T. disciformis*), G2M, group 2 (*T. eikelboomii*) and G3M, group 3 (*T. flexilis*).

We recommend using probe G123T and its competitor probe to detect the entire *Thiothrix*/021N group. As these probes may target filaments other than *Thiothrix*/021N, simultaneous application of probe G123T with probes TNI and 21N is recommended (Schade M., unpublished). For further differentiation between these type 021N filaments (*T. eikelboomii, T. flexilis* and *T. disciformis*) and *Thiothrix* species more specific probes can be used, refer to Figure 5.7. Also the class-specific probe for *Gammaproteobacteria* targets these morphotypes (Table 5.4). See also FISH images in Figures 9.40–9.44, Chapter 9.

5.3.10 0041/0675 morphotype

Abundance, morphology and affiliation

This morphotype has been identified in WWTPs all over the world. 0041/0675 filaments are ranked as the second to sixth most frequently seen bacteria (summarized in Jenkins *et al.*, 2004; Tandoi *et al.*, 2006). FISH surveys have also been published with a selection of probes targeting the 0041/0675 morphotype. For example, approximately 15% of all morphotype 0041/0675 filaments were targeted by the probe for TM7 subdivision 1 in Danish WWTPs (Thomsen *et al.*, 2002). A German survey detected a larger fraction (approx. 40%) of TM7 affiliated morphotypes (Müller *et al.*, 2007).

The 0041/0675 morphotype has straight or bent filaments with a cell diameter of 0.7 to 2 μm and almost square cells. Filaments often show attached growth. They stain mainly Gram positive and Neisser negative. A faint Neisser staining response might suggest phosphorus deficiency in the plant.

The phylogenetic affiliation of members of morphotype 0041/0675 is diverse. Many show affiliation with members of the *Chloroflexi* (Bjornsson *et al.*, 2002; Kragelund *et al.*, 2007a). Others affiliate with *Curvibacter* (recently reclassified from *Aquaspirillum*) within the *Betaproteobacteria* (Thomsen *et al.*, 2006b) and some are embraced by the Candidate division TM7 (Hugenholtz *et al.*, 2001). However, blast searches of the TM7 probes show that approx. 50% of the targeted sequences are in fact affiliated to the phylum of *Chloroflexi.* Therefore, a thorough cross check of probe results with the filament morphological details after phase contrast microscopy is highly recommended.

Recommending the best probes

Table 5.11 shows the oligonucleotide probes available for identifying and quantifying morphotype 0041/0675, while the tree in Figure 5.8 illustrates their specificity and coverage.

Morphotype 0041/0675 filaments are most often identified with the phylum-targeted CFXmix probes for the *Chloroflexi* (Bjornsson *et al.*, 2002; Kragelund *et al.*, 2007a). These also detect morphotypes 0092 and 1851 and some *H. hydrossis*-like bacteria (see other sections). For more precise identification, genus- and species-targeted probes within the *Chloroflexi* phylum are required, but these are currently not available. Some 0041/0675 morphotypes hybridize with probe Curvi997 when applied with competitor probes Comp1curvi997 and Comp2curvi997 (Thomsen *et al.*, 2006b) and thus belong to the genus *Curvibacter.* However, this probe also targets some abundant microcolony-forming bacteria present in many WWTPs (Thomsen *et al.*, 2004 and Chapter 3) and in addition hybridizes with type 1701. Some 0041/0675 filaments are detected with probes TM7905 targeting almost the entire Candidate division TM7, and TM7305, which target members of subdivision 1. No taxonomically defined species within this phylum have been assigned so far to type 0041/0675 filaments. In addition, single cells and other filaments have been seen hybridizing with these TM7905 and TM7305 probes (Hugenholtz *et al.*, 2001).

We recommend applying probes for each phylum in studies with type 0041/0675 i.e. probes CFX mix for *Chloroflexi* members, Curvi997 and competitor probes for some *Betaproteobacteria* and probes TM7905 and TM7305 for those in candidate division TM7. Remember to cross-check the FISH data with those obtained with phase contrast microscopy. See also FISH images in Figures 9.45–9.48, Chapter 9.

Table 5.11. Morphotype 0041/0675: rRNA-targeted oligonucleotide probes with corresponding formamide concentration, target, morphotype, and comments on recommendability for bacteria with attached growth.

Probe name	Target	Morphotype	Sequence (5′-3′)	FA (%)	Comments	Reference
CFX mix (GNSB941 +CFX1223)	Phylum *Chloroflexi*	Type 0041/0675	AAACCACACGCTCCGCT+ CCATTGTAGCGTGTGTGTMG	35	Also other morphotypes can be observed, often type 1851 and sometimes also type 1701, and *H. hydrossis*-like filaments	Bjornsson *et al.* 2002 Gich *et al.* 2001
TM7305	subdivision 1 of candidate division TM7	Type 0041/0675	GTC CCA GTC TGG CTG ATC	30	Primarily type 0041/0675	Hugenholtz *et al.* 2001
Curvi997 Comp1curvi997+ Comp2curvi997	*Curvibacter delicatum* ATCC14667; *Pseudomonas lanceolata* AB021390 + few more (*Betaproteobacteria*) Competitor to curvi997 should be applied together	Type 0041/0675	CTC TGG TAA CTT CCG TAC CTC TGG CAA CTT CCG TAC + CTC TGG TCA CTT CCG TAC	35	Primarily targets type 1701 and microcolony forming bacteria therefore please check thoroughly for morphology consistency	Thomsen *et al.* 2004

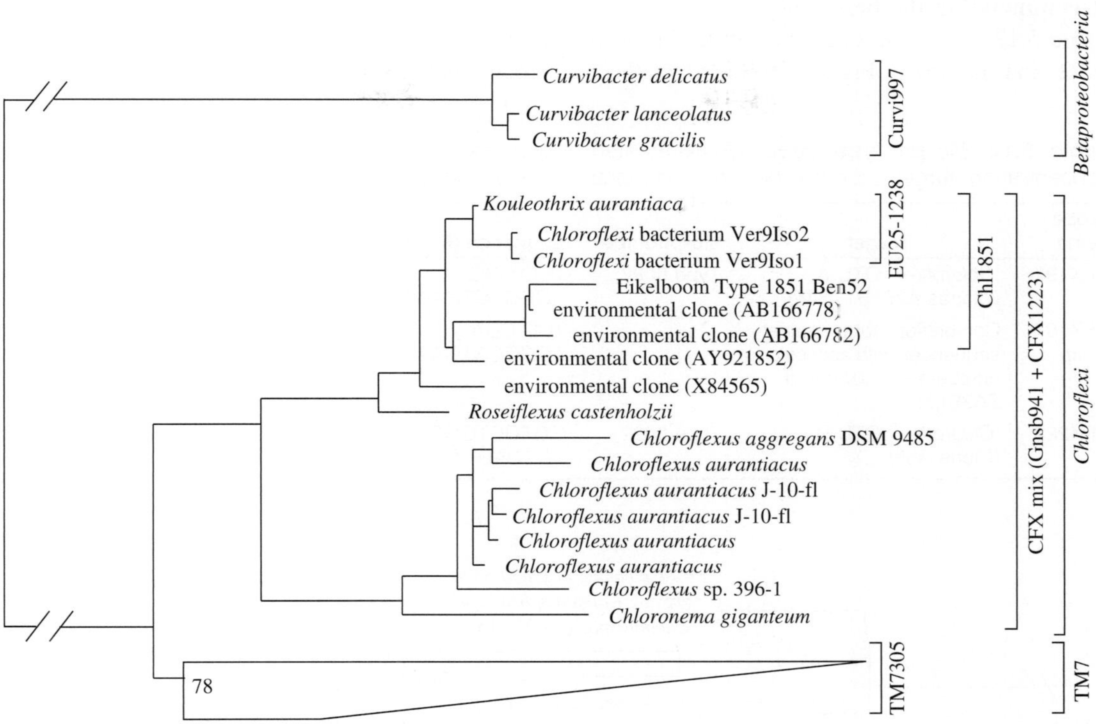

Figure 5.8. 16S rRNA-based phylogenetic tree showing main lineages of known bacteria with attached growth. Brackets indicate the coverage of the probes listed in Tables 5.11 and 5.13–5.14. The exact branching order of the bacteria with attached growth lineages cannot yet be determined due to a limited phylogenetic resolution of the 16S rRNA marker gene. Therefore, unclear tree topology is depicted as multifurcation.

5.3.11 0092 morphotype

Abundance, morphology and affiliation

Type 0092 appears prominently in microscopy-based surveys of activated sludge plants around the world (Jenkins *et al.*, 2004). An Australian FISH survey of EBPR plants where this morphotype was abundant, showed that it never responded to the EUBmix probes (Speirs *et al.*, 2009). Such survey data suggest that it seems to be associated especially with plants operating with low F/M feeds and long sludge ages, and thus is seen frequently in activated sludge systems removing P. However, it does not stain positively for polyP granules with DAPI.

Its morphotype is distinctive, appearing as short blunt ended trichomes extruding from flocs or occasionally freely suspended in the bulk liquid. Some reports suggest it is also enriched in foams, although whether it is a causative organism is unlikely. It can be 'identified' readily microscopically by its Neisser positive staining reaction, where filaments appear distinctively 'lilac' in color. It stains Gram negative.

Reports that this morphotype has been cultured should be treated cautiously. The phylogeny of most isolates has never been resolved, and where 16S rRNA sequencing has been performed, FISH probes designed against them have never hybridized with this filament morphotype *in situ*.

Recommending the best probes

Table 5.12 shows the oligonucleotides probes recommended now for identifying and quantifying type 0092, and the tree in Figure 5.9 illustrates their specificity and coverage.

Table 5.12. Morphotype 0092: rRNA-targeted oligonucleotide probes with corresponding formamide concentration, target, morphotype, and comments on recommendability.

Probe name	Target	Morphotype	Sequence (5′-3′)	FA (%)	Comments	Reference
CFX197	'*Chloroflexi*' OTU A (Clones A26, B1, B45)	Type 0092	TCC CGG AGCGCCTGAACT	40		Speirs *et al.*, 2009
CFX197 comp	Competitor probe against sequences with accession numbers ZA3635c and ZA3612c		TCC CGA AGCGCCTGAACT			Speirs *et al.*, 2009
CFX223	'*Chloroflexi*' OTU B (Clone A58)	Type 0092	GGTGCTGGCTCC TCCCAG	35		Speirs *et al.*, 2009

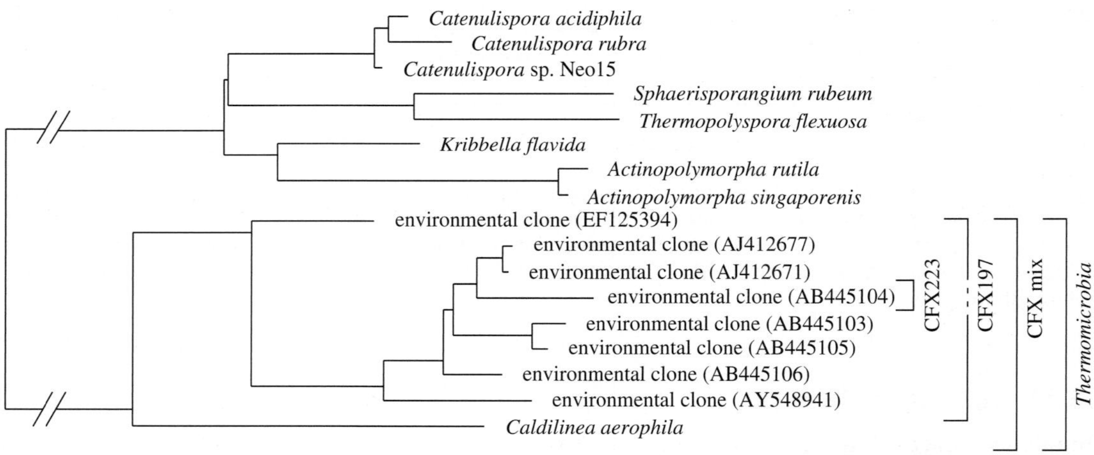

Figure 5.9. 16S rRNA-based phylogenetic tree showing main lineages of morphotype 0092 belonging to the *Chloroflexi*. Brackets indicate the coverage of the probes listed in Table 5.12. The exact branching order of the *Chloroflexi* lineages cannot yet be determined due to a limited phylogenetic resolution of the 16S rRNA marker gene. Therefore, unclear tree topology is depicted as multifurcation.

Morphotype 0092 hybridized with the GNSB941 and CFX1223 probes targeting the *Chloroflexi*, but not the CFX109 or CFX784 probes for *Chloroflexi* subgroups 3 and 1a, or CHL1851 designed against the *Chloroflexi* type 1851 filament. Two more specific 16S rRNA-targeted probes were designed against clone sequences recovered from one plant (Speirs *et al.*, 2009). The CFX197 probe hybridized with the slightly thicker of two morphological variants of this morphotype in several Australian P-removal plant samples, while CFX223 probe targeted the thinner variant. Is is not clear whether these two variants represent different species or strains of this filament. With all FISH analyses, the fluorescent signal from type 0092 was localized, giving the filaments a granular appearance, suggesting that their ribosomes are in aggregates in cells.

We recommend applying the probes CFX197 and CFX223 probes for targeting morphotype 0092. However, the probes need to be applied to more biomass samples from plants around the world before this issue can be addressed confidently. However, on present evidence, type 0092 morphotypes should be screened initially with the EUBmix, GNSB941 and CFX1223 probes, and then if negative for the first and positive for the latter two probes, the CFX197 and CFX223 probes should be used. Whether they hybridize globally with morphologically different type 0092 filaments is of interest. See also FISH images in Figures 9.49–9.51, Chapter 9.

5.3.12 1701 morphotype

Abundance, morphology and affiliation

Type 1701 is only seen occasionally, particularly in US plants, and is considered to cause severe bulking problems only rarely (Jenkins *et al.*, 1993; Thomsen *et al.*, 2006b). Filaments occur often in bundles.

The 1701 morphotype forms short filaments which are often attached to flocs. The trichome has a diameter of 0.7–1.1 μm, and consists of rod-shaped cells. Attached growth is often present. It is usually Gram and Neisser negative.

The identity of this morphotype is unclear. *S. natans* was suggested as a close relative (Howarth *et al.*, 1998) based on 16S rRNA gene sequence analyses, and it is placed within the genus *Sphaerotilus*. Wagner *et al.* (1994a) also found type 1701 was closely affiliated with *Sphaerotilus* spp. and was readily detected by FISH with the probe SNA in full-scale plants. Other strains of this morphotype belong possibly to *Aquaspirillum delicatum* within the *Betaproteobacteria* (Thomsen *et al.*, 2006b), which is now classified as *Curvibacter delicatus* (Ding and Yokota, 2004), a notation used throughout this handbook.

As stressed with other filamentous bacteria with attached growth (types 0041/0675 and 1851) a thorough cross-check of FISH images with those from phase contrast microscopy is recommended.

Recommending the best probes

Table 5.13 shows the oligonucleotides probes for identifying and quantifying type 1701, and the tree in Figure 5.8 illustrates their specificity and coverage.

The probe Curvi997 (formerly Aqs997) with competitor probes Comp1curvi997/Comp2curvi997 (formerly Comp1aqs997/Comp2aqs997) were developed initially for microcolony-forming bacteria. Curvi997 was shown subsequently to target bacteria with the type 1701 morphology (Thomsen *et al.*, 2004). Two variants were observed with this probe: one resembles type 1701 and the other is more similar to type 0041/0675. The 1701 morphotypes are frequently observed in low loaded plants with N removal and biological P removal but only occasionally cause bulking there (Thomsen *et al.*, 2006b).

We recommend applying Curvi997 and respective competitor probes for the *Curvibacter*-related type 1701. However, this probe targets other filament morphotypes resembling type 0041. As type 1701 is easily confused with several other morphotypes, its staining behavior (1701 Gram−, 0041/0675 Gram+) and morphological features under phase contrast should first be checked carefully (e.g. rod-shaped cells in short filaments (1701) or square cells in long filaments (0041). In addition probe SNA should also be applied for type 1701 strains affiliated with *Sphaerotilus* spp. (Wagner *et al.*, 1994a). See also FISH images in Figures 9.52–9.53, Chapter 9.

Table 5.13. Morphotype 1701: rRNA-targeted oligonucleotide probes with corresponding formamide concentration, target, morphotype, and comments on recommendability for bacteria with attached growth.

Probe name	Target	Morphotype	Sequence (5′-3′)	FA (%)	Comments	Reference
Curvi997 Comp1curvi997+ Comp2curvi997	*Curvibacter delicatum* ATCC14667; *Pseudomonas lanceolata* AB021390 + few more. Competitors to curvi997 should be applied together	Type 1701	CTC TGG TAA CTT CCG TAC CTC TGG CAA CTT CCG TAC + CTC TGG TCA CTT CCG TAC	35	Sometimes also type 0041/0675 and microcolony-forming bacteria	Thomsen *et al.* 2004
CFX mix (GNSB941 +CFX1223)	Phylum *Chloroflexi*	Type 1701	AAACCACACGCTCCGCT+ CCATTGTAGCGTGTGTGTMG	35	Primarily type 0041/0675 and type 1851 and sometimes also *H. hydrossis*- like filaments	Bjornsson *et al.* 2002 Gich *et al.* 2001
SNA	*Sphaerotilus natans* (*Betaproteobacteria*)	Type 1701 *S. natans* *Leptothrix* spp.	CAT CCC CCT CTA CCG TAC	45		Wagner *et al.* 1994a

5.3.13 1851 morphotype

Abundance, morphology and affiliation

The 1851 morphotype is commonly seen in WWTPs all over the world, although not associated with bulking incidences in the US (summarized in Jenkins *et al.*, 2004). It is found in municipal and industrial WWTPs with and without nutrient removal (Eikelboom, 2000; Beer *et al.*, 2002; Jenkins *et al.*, 2004; Kragelund *et al.*, 2007a), and is responsible occasionally for severe bulking and foaming incidents in German municipal low F/M WWTPs (Lemmer *et al.*, 2000). FISH surveys have also been conducted using probes targeting morphotype 1851, where long straight to slightly bent filaments forming twisted bundles were detected predominantly in municipal WWTPs in Australia, New Zealand, the Netherlands, and France (Beer *et al.*, 2002), Japan (Kohno *et al.*, 2002), and in German WWTPs with scumming events (Müller *et al.*, 2007). Morphotype 1851 has also been observed in plants with high industrial loads and can be observed both with and without attached growth (Kragelund *et al.*, 2007a).

This morphotype shows straight or slightly bent long filaments that often form characteristic bundles. Its cell diameter is between 0.6 and 0.8 μm. Attached growth is nearly always present in filaments seen in domestic plants, but is often missing in those from industrial WWTPs. It stains Gram positive, characteristically as light blue, and is usually Neisser negative.

Several morphotype 1851 strains have been identified as *Kouleothrix aurantiaca* in the phylum *Chloroflexi* (class *Roseiflexus*) after 16S rRNA sequence analysis of micromanipulated filaments from activated sludge samples (Beer *et al.*, 2002; Kohno *et al.*, 2002; Kragelund *et al.*, 2007a).

Recommending the best probes

Table 5.14 shows the oligonucleotide probes for identifying morphotype 1851, and the tree in Figure 5.8 illustrates their specificity and coverage.

Two probes (CHL1851 and EU25-1238) were designed to target this morphotype in activated sludge (Beer *et al.*, 2002; Kragelund *et al.*, 2007a). A survey of samples from industrial WWTPs showed that approx. 50% of probe EU25-1238 positive filaments did not hybridize with the probe CHL1851 (Kragelund *et al.*, 2007a), even though the EU25 isolate possessed the probe target site at *E. coli* position 592.

Morphotype 1851 strains from predominantly municipal WWTPs are readily detected with the probe CHL1851. However, we recommend using probe EU25-1238 for detecting this morphotype in WWTPs with both high and low loadings of industrial wastewater. Many of the EU25-1238 probe-positive filamentous bacteria in industrial WWTPs have little or no attached growth. See FISH images in Figures 9.54–9.56 in Chapter 9.

5.3.14 1863 morphotype

Abundance, morphology and affiliation

Morphotype 1863 is observed in WWTPs only occasionally and seems to play no important role in bulking (Jenkins *et al.*, 2004; van der Waarde *et al.*, 2002). Type 1863 and unicellular *Acinetobacter* cells can be detected in some foams or scums since these produce often hydrophobic capsular material and are known to release biosurfactants under nutrient limitation (Lemmer *et al.*, 2000; Müller *et al.*, 2007). In a survey of municipal WWTPs 60% contained Gram negative coccoid cells and 30% had type 1863 filaments which were detected by the probe ACA23a (Wagner *et al.*, 1994b) specific for the morphotype 1863 affiliating with the genus *Acinetobacter* in the *Gammaproteobacteria* (Müller *et al.*, 2007). However, it also targets straight filaments with square-shaped cells (Schade M., unpublished results). When this probe was applied to samples from industrial WWTPs (van der Waarde *et al.*, 2002), the morphotype 1863 was detected in approx. 16% of plants. Their relative abundances were not reported (van der Waarde *et al.*, 2002).

Table 5.14. Morphotype 1851: rRNA-targeted oligonucleotide probes with corresponding formamide concentration, target, morphotype, and comments on recommendability for bacteria with attached growth.

Probe name	Target	Morphotype	Sequence (5′-3′)	FA (%)	Comments	Reference
CFX mix (GNSB941 +CFX1223)	Phylum *Chloroflexi*	Type 1851	AAACCACACGCTCCGCT+ CCATTGTAGCGTGTGTGTMG	35	Also other morphotypes can be observed, often type 0041/0675 and sometimes also type 1701, and *H. hydrossis*-like filaments	Bjornsson *et al.* 2002 Gich *et al.* 2001
CFX 109	Some members of the class *Chloroflexi*	Type 1851	CACGTGTTCCTCAGCCGT	30	Also other morphotypes can be observed, often type 0041/0675 and sometimes also Type 1701, and *H. hydrossis*-like filaments	Bjornsson *et al.* 2002 Gich *et al.* 2001
EU25-1238	Isolate EU25 and other *Kouleothrix aurantiaca* (class *Chloroflexi*)	Type1851	CTG CGC ATT GCC ACC GAC AT	35	Type 1851 also without epiphytic growth, predominantly industrial	Kragelund *et al.* 2007
CHL1851	Type 1851 *Kouleothrix aurantiaca* (class *Chloroflexi*)	Type 1851	AAT TCC ACG AAC CTC TGC CA	20	Type 1851 with and without epiphytic growth, predominantly municipal	Beer *et al.* 2002

However, another survey where the abundance of morphotype 1863 in 126 industrial WWTPs was reported, only 4% of the plants contained this morphotype and the filament index never exceeded 1 (Kragelund C and Nielsen PH, unpublished results).

Type 1863 is characterized by bent, curled or tangled short filaments consisting of regular cells of 0.5 to 0.7 μm diameter. Cells are spherical to ovoid, and stain Gram and Neisser negative.

The morphotype 1863 is polyphyletic. Some isolates affiliate with members of the genus *Acinetobacter* in the *Gammaproteobacteria* (also targeted by class-specific probe GAM42a, Table 5.4) (Wagner *et al.*, 1994b), while others affiliate with *Chryseobacterium* spp. within the *Bacteroidetes* (Seviour *et al.*, 1997). It can also be targeted by probe CFB563, Table 5.4.

Recommending the best probes

Table 5.6 shows the oligonucleotide probes for identifying and quantifying morphotype 1863, and the tree in Figure 5.3 illustrates their specificity and coverage.

The probe ACA23a was designed to detect *Acinetobacter* spp. in the *Gammaproteobacteria* in activated sludge (Wagner *et al.*, 1994b). Applied to municipal activated sludge samples ACA23a visualizes rods and/or cocci in clusters and chains (Wagner *et al.*, 1994b), see FISH image Figure 9.57 in Chapter 9. No probes have been designed for the type 1863 affiliating with *Chryseobacterium* spp. within the *Bacteroidetes.*

Morphotype 1863 never causes severe bulking problems. As it can be involved in scum formation we recommend that the ACA23a probe for *Acinetobacter*-related 1863 strains be used in such cases.

5.3.15 0803, 0914 morphotypes and other still unidentified filamentous species

Abundance, morphology and affiliation

A number of morphotypes are still currently unidentified (e.g. morphotype 0211, 0411, 0581, 0803, 0914, 0961, 1702 and 1802). However, many of these (morphotypes 0211, 0411, 0581, 0961, 1702 and 1802) are only rarely observed and therefore not considered important enough for inclusion in this manual. Some morphotypes observed more often in activated sludge like morphotypes 0803 and 0914 will be discussed.

Morphotypes 0803, and 0914 have been 'identified' by microscopy in activated sludge communities around the world, and are often seen in large numbers. Thus, morphotype 0803 was ranked as the fourth most abundant filament in Czech Republic, and morphotype 0914 ranked fifth (surveys summarized in Jenkins *et al.*, 2004 and Tandoi *et al.*, 2006).

Both type 0803 and 0914 form short (<50 μm) straight, or longer (100–300 μm) bent filaments. Filaments are attached sometimes by a holdfast to inorganic particles. The cell diameter is between 0.8 and 1.1 μm. Cells are mostly square (0914) or rectangular (0803). With sulfur stores, type 0803 becomes difficult to distinguish from type 0914 according to the descriptions of Eikelboom (2000). However, different cell shape and cell diameter are found *in situ* and can be used for their differentiation. These filaments can be Gram variable, but are more usually Gram negative, and occasionally feature Neisser positive granules.

The phylogenetic affiliation is still not resolved for any of these morphotypes. Partial 16S rRNA gene sequence analysis from a micromanipulated morphotype 0803 isolate showed an affiliation with the *Betaproteobacteria* (Bradford *et al.*, 1996). However, no *in situ* hybridization with the general betaproteobacterial probe BET42a is observed (Kragelund C, unpublished results). So the phylogenetic affiliation of type 0803 is still uncertain. Morphotype 0914 is still unidentified and to our knowledge no successful attempts have been made to sequence the 16S rRNA gene (either from an isolate or from micromanipulated filaments). However, very recent results indicate that they may belong to the *Chloroflexi*, hybridizing with the probes GNSB941 and CLX 1223 (Seviour R, unpublished) and new probes are presently being developed.

6

Identification of other microorganisms in activated sludge and biofilms by FISH

Per Halkjær Nielsen and Jeppe Lund Nielsen

6.1 INTRODUCTION

Besides the functional groups described in previous chapters, many other groups of bacteria are present in wastewater treatment systems, although often in low numbers. In this chapter we have included a few groups that are present or important in many plants, they may have an important function, and relatively specific gene probes are available. However, we do not provide detailed recommendations, only up-to date references.

6.2 EPIPHYTIC BACTERIA INVOLVED IN PROTEIN HYDROLYSIS

In full-scale treatment plants several types of filamentous bacteria are often observed with a large number of epiphytic, attached bacteria. These are known as type 0041, type 1851, and type 1701 based on morphological identification (see Chapter 5 about filamentous bacteria). Most of these filamentous bacteria belong to the candidate phylum TM7, the phylum *Chloroflexi*, and the class *Betaproteobacteria* (related to *Curvibacter*) and can be detected by gene probes. Recently, it has been shown that most of their epiphytic bacteria belong to the *Saprospiraceae* in the phylum *Bacteroidetes* (Kong *et al.*, 2007; Xia *et al.*, 2008). A new genus *Candidatus* Epiflobacter was proposed for this epiphytic group. Three clusters were identified but no differences in physiology could be detected. These bacteria are specialized in protein hydrolysis and use amino acids as energy and carbon sources. They are not involved in denitrification and cannot store polyphosphate or PHA. These epiphytic rods are typically perpendicularly attached onto the filaments and are 1.3–1.9 μm in length and 0.3–0.4 μm in width. They can be very abundant in BNR activated sludge treatment plants and account up to 10% of all bacterial biovolume.

Most of the epiphytic bacteria hybridize with probe SAP-309 targeting *Saprospiraceae* in the phylum *Bacteroidetes*, see Chapter 5. Probe Bac111 (Kong *et al.,* 2007) is more specific for species found in wastewater treatment plants and targets also most of these epiphytic bacteria. Three probes exist for detection of three subgroups but as no differences in abundance or ecophysiology is known today, it is recommended to apply probe Bac111 to detect almost all epiphytic *Saprospiraceae* (Kong *et al.,* 2007; Xia *et al.,* 2008).

6.3 SULFATE-REDUCING BACTERIA

Sulfate-reducing bacteria are present in most activated sludge systems (Lens *et al.,* 1995), and are physiologically active during anaerobic periods, for example, in the anaerobic reactors of EBPR plants or in sludge storage tanks (Rasmussen *et al.,* 1994; Kjeldsen *et al.,* 2004). Prolonged aeration may not lower their activity, and oxygen-tolerant sulfate-reducing *Desulfovibrio* strains isolated from activated sludge plants have been described (Kjeldsen *et al.,* 2005; Mogensen *et al.,* 2006). About 0.5–8% of the total bacteria have been identified as sulfate-reducers in full-scale WWTPs (Manz *et al.,* 1998; Vester and Ingvorsen, 1998; Nielsen and Nielsen, 2002a). The dominant SRB belonged to the families *Desulfovibrionaceae* and *Desulfobacteraceae* (Manz *et al.,* 1998).

Sulfate-reducing bacteria are a polyphyletic group with most known members in the *Deltaproteobacteria*. Their detection by FISH will not be described in detail in this book and the readers are recommended to use the studies by Manz *et al.* (1998) and Lücker *et al.* (2007).

6.4 FERMENTING BACTERIA

Fermentation in activated sludge has been shown to be performed partially by bacteria belonging to the genus *Streptococcus* in the class *Bacilli* and the family *Streptococcaceae* (Kong *et al.,* 2008). The physiology of the fermenting streptococci reveals a broad group of cells which are capable of partially oxidizing glucose to produce lactate, acetate, ethanol, and formate (Liu *et al.,* 2002; Kong *et al.,* 2008). These are facultative anaerobes, utilizing glucose under both aerobic and anaerobic conditions (but not mannose or galactose) (Nielsen PH, unpublished). They probably ferment substrates in the anaerobic reactors, thus being important in providing readily metabolizable substrates for the denitrifying and phosphate-removing communities in these processes. Streptococci are quite abundant in many municipal WWTPs with EBPR, constituting 1–4% of the biovolume (Kong *et al.,* 2008). Some of these are filamentous, see Chapter 5.

The probe-defined consortia hybridizing with probe Str/Strept (Trebesius *et al.,* 2000) can be used to identify fermenting *Streptococci* (see also Chapter 5). The probe targets 44% of all streptococci, and can be used together with the broader probe LGC354C (Meier *et al.,* 1999). The morphologies of the Str/Strept-probed cells cover both small cocci and filamentous organisms. A broader probe, Strc498, targets most *Streptococcus* spp. (Franks *et al.,* 1998), but has to our knowledge not been tested in wastewater systems.

6.5 *ESCHERICHIA COLI* AS INDICATOR ORGANISM FOR ENTERO-PATHOGENS

Fecal material from human and animals contain a large variety of pathogenic microorganisms. A systematic search of all potential strains of entero-pathogens is not feasible, thus enumeration of gammaproteobacterial *Escherichia coli* has been used as indicator bacteria to evaluate the fecal content and the sanitary risk. Several attempts of using FISH to enumerate *E. coli* in municipal WWTPs have been performed. The need for accurate numbers of potentially active *E. coli* in natural waters has resulted in approaches by which *E. coli* has to be revived prior to FISH (Garcia-Armisen and Servais, 2004). Direct viable count (DVC) procedures involve exposing the cells to a revivification medium containing antibiotics preventing cellular division; elongated cells are then enumerated as viable cells.

Using such DVC-FISH approaches typically around 1% of the total cell number appear as *E. coli* (Garcia-Armisen and Servais, 2004), whereas very few studies have been able to enumerate the specific pathogens for which *E. coli* function as indicator (e.g. *Campylobacter jejuni, Clostridium perfringens, Bacillus cereus, Enterococcus faecalis, E. coli* O157:H7, *Helicobacter pylori, Klebsiella pneumoniae, Legionella pneumophila, Listeria monocytogenes, Pseudomonas aeruginosa, Salmonella* spp. and *Staphylococcus aureus*). However, from quantitative PCR these have been found to estimate several orders below that of *E. coli* (e.g. Shannon *et al.,* 2007).

Escherichia in the activated sludge is believed to be coming from the incoming wastewater. To what extent these survive or even grow here is not known. The appearance under these conditions is rods, typically 1.0–2.3 μm in length and 0.4–0.5 μm in width.

Several probes have been used for the identification and quantification of *E. coli*. The probe colinsitu has a central mismatch towards the target site in the genera *Escherichia* and *Shigella* both within the family *Enterobacteriaceae*, but colinsitu has fewer outgroups hits compared to the perfect match probe Ec637 (Regnault *et al.,* 2000). Since both probes yield good FISH signals the colinsitu probe may be recommended for detection of *E. coli*.

7

Protocol for Fluorescence *in situ* Hybridization (FISH) with rRNA-targeted oligonucleotides

Jeppe Lund Nielsen

7.1 INTRODUCTION

Fluorescence *in situ* hybridization (FISH) using rRNA-targeted oligonucleotide probes has become one of the most widely used approaches when studying microorganisms directly in complex systems without prior cultivation and isolation. The method makes it possible within a relatively short time to retrieve information on phylogenetic identities of the cells directly in a sample, and since it also maintains the morphology of the cells it also supplies information on the spatial distribution as well as the number of identified organisms.

The first applications of *in situ* hybridization were applied using radioactively labeled oligonucleotides, which had limited microscopic resolution and needed autoradiography for visualization. The use of FISH with fluorescently labeled probes was first described in 1989 (DeLong *et al.*, 1989) and has since become the method of choice for examining the microbial consortia in both natural and engineered systems. In wastewater microbiology, the FISH approach has been shown to be especially well-suited with high cell signals and detection sensitivities.

The principle of the FISH technique is based on hybridizing fluorescently labeled probes to ribosomal rRNA in permeabilized whole microbial cells. The probes consist of short pieces of DNA (usually 15–25 nucleotides in length) and they are designed to specifically hybridize to their complementary target sequence on the rRNA structures (16S and 23S subunits are typically used for *Bacteria*, while for *Eucarya* the 18S subunit is the most used; Amann, 1995) in the target cell. From the composition of the probe, it is possible to design it to specifically target a narrow phylogenetic group (down to the species level) or any other higher phylogenetic hierarchical group (top to bottom approach, Amann *et al.*, 1995). No probes will hybridize to those cells without target sequences. Cells containing the target sequence will on the other hand retain the hybridized probe and due to the large number of ribosomes within active cells thus become fluorescently labeled.

Several recent reviews have elegantly described the potentials of the FISH technique and new developments for improved sensitivity (e.g. Wagner *et al.*, 2003; Zwirglmaier, 2005; Bottari *et al.*, 2006; Amann and Fuchs, 2008). Often occurring pitfalls within the application of the FISH technique and solutions hereto have been reviewed by Moter and Göbel (2000), while important factors that have an influence on the sensitivity and quality of the obtained data are described by Bouvier and del Giorgio (2003).

The main obstacles associated with the FISH method are poor cell permeability, insufficient ribosome content, ribosome inaccessibility, and sample autofluorescence. Thus knowledge of the nature and applicability of the sample as well as a uniform protocol with application of the proper controls are of fundamental importance for obtaining solid and comparable information.

This chapter provides a detailed protocol for applying FISH on activated sludge samples and possible solutions to typical problems encountered. A typical flowchart for FISH in activated sludge or biofilm is shown in Figure 7.1 and includes sampling, possible pretreatment, fixation, possible enzyme treatment, hybridization, epifluorescence microscopy, and possible image analysis. Details are described below.

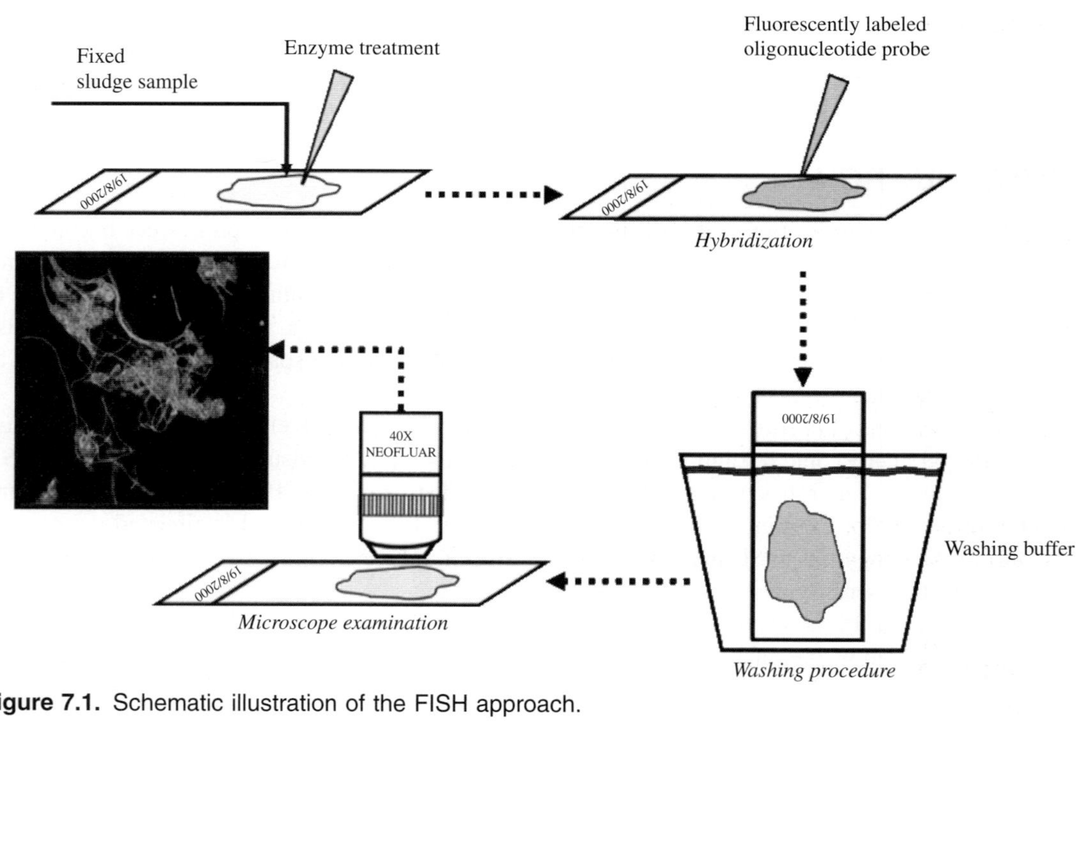

Figure 7.1. Schematic illustration of the FISH approach.

7.2 FISH PROTOCOL

7.2.1 Materials and solutions

8% Paraformaldehyde/PBS (8% PFA)
Stir 4 g PFA in 30 mL dH_2O at 60°C and dissolve with a drop of 2 N NaOH (in the fume hood!); add 16.6 mL 3× phosphate-buffered saline (PBS) (see below). Adjust with dH_2O to 50 mL. The solution is filtrated through a 0.22 μm polycarbonate filter to remove autofluorescing particles. Should be used fresh, or stored in aliquots at −20°C (do not freeze thawed solutions).

1:1 PBS/Ethanol (PBS/EtOH)
Mix equal volumes of 1×PBS and 96% ethanol, sterilize by filtration.

Gelatine solution
Dissolve 0.5 g gelatine and 0.01 g $CrK(SO_4)_2$ in 100 mL dH_2O at 70°C. Store at 4°C.

Poly-L-lysine solution
Dissolve 0.01 g poly-L-lysine in 100 mL dH_2O.

Lysozyme
Dissolve lysozyme (Merck Chemicals, Denmark) to a final concentration of 10 mg/mL (~360 000 U/mL) - in 0.05 M EDTA, 0.1 M Tris-HCl, pH 8.0 (prepare fresh when needed; keep on ice until use).

Mutanolysin
Dissolve mutanolysin (Sigma-Aldrich, St. Louis, MO, USA) to 5000 U/mL in 0.1 M potassium phosphate buffer, pH 6.2 (prepare fresh when needed; keep on ice until use).

Achromopeptidase
Prepare achromopeptidase (Sigma-Aldrich, St. Louis, MO, USA) solution to a final concentration of 60 U/mL in 0.01 M NaCl, 0.01 M Tris-HCl (pH 8.0) (prepare fresh when needed, keep on ice until use), alternatively, prepare aliquots in tubes and freeze until needed.

Proteinase K
Dissolve proteinase K (*Tritirachium album*, Sigma-Aldrich, St. Louis, MO, USA) to 20 000 U/mL in TE buffer (0.01 M EDTA, 0.1 M Tris-HCl, pH 8.0). (Prepare fresh when needed; keep on ice until use).

Lipase
Dissolve lipase (Sigma-Aldrich, St. Louis, MO, USA) to 75 000 U/mL in 1×PBS (prepare as 3×PBS and dilute with dH_2O) (Prepare fresh when needed; keep on ice until use).

3× Phosphate-buffered Saline (3×PBS)
Mix 0.1 M NaH_2PO_4 with 0.1 M Na_2HPO_4 till pH 7.4. Use 300 mL hereof and mix with 22.8 g NaCl and adjust to 1000 mL with dH_2O. Autoclave, and store at room temperature.

Tris-EDTA buffer (TE buffer)
0.01 M EDTA (ethylenediaminetetraaceticacid), 0.1 M Tris-HCl, pH 8.0. Sterilize by filtration and store at 4°C.

1 M Tris-HCl, pH 8.0
Dissolve 121.1 g Tris in 800 ml dH_2O, add 42 mL concentrated HCl, let cool, adjust pH and fill to 1L with dH_2O. Autoclave, and store at room temperature.

5 M NaCl
Dissolve 292.2 g NaCl in 800 mL dH_2O, fill to 1 L with dH_2O. Sterilize by filtration, and store at room temperature.

Sterile distilled H_2O (dH_2O)
Autoclave sterile filtered distilled H_2O.

10% Sodium dodecylsulfate (10% SDS)
Heat 50 g SDS (electrophoresis-quality) in 400 mL dH_2O to 70°C, adjust pH with concentrated HCl to 7.2, fill to 500 mL; no sterilization required. Store at room temperature.

0.5 M EDTA
Dissolve 18.6 g EDTA in 80 mL dH_2O by adjusting the pH to 8.0 (ca. 2 g NaOH pellets required), fill to 100 mL with dH_2O. Sterilize by filtration and store at 4°C.

7.2.2 Equipment and supplies needed for FISH

- Water bath (46°C)
- Hybridization oven (48°C)
- Epifluorescence microscope or Confocal Laser Scanning Microscope (CLSM) equipped with the proper filters
- Tissue grinder (Thomas Scientific®, Swedesboro, NJ, USA)
- Citifluor AF1 (Citifluor Ldt, London, England)
- Vectashield (VECTASHIELD® Hard-Set™ Mounting Medium, Vector Laboratories, Inc., Burlingame, CA, USA)
- Teflon-coated glass slides separating 6 or 10 reaction fields = wells (Paul Marienfeld GmbH & Co, Lauda-Königshofen, Germany)
- Tissue-Tek O.C.T. (Sakura Finetek, Zoeterwoude, The Netherlands)

7.3 PROTOCOL

7.3.1 Sample collection and fixation

Fresh samples are collected and fixed as soon as possible (activated sludge and biofilms can usually be kept in the refrigerator 2–4°C for 2–3 days without major changes).

Sample preparation prior to hybridization with the fluorescently labeled probes involves fixation and cell wall permeabilization in order to inactivate the microbial cells and any enzymatic activity, to avoid growth/decay after harvest and to permeabilize the cells for probe penetration. The most commonly used fixative agent is polymerized formaldehyde (paraformaldehyde) which preserves the cell morphology, although some Gram positive cells are not adequately permeabilized with this fixative. Thus the sample is usually fixed independently for both Gram negative and Gram positive cells.

Gram positive cells:
Mix equal volumes of sludge and 96% ethanol in a test tube. Keep the solution in the freezer (−20°C).

Gram negative cells:
Mix equal volumes of sludge with cold 8% PFA.

Incubate on ice for ½–3 hours, depending on sample texture and target organisms (some cells require short incubations in order to allow permeabilization of the probe, while other cells remain FISH-detectable even after several hours of fixation. Typically sludge or biofilm samples with flocculated material must be fixed for 2–3 hours to allow an even fixation of interior cells in the flocs/biofilms).

Separate PFA and biomass by centrifugation (3500 g for 8 min).

Dissolve the pellet in 5 mL cold 1:1 PBS/EtOH. Separate by centrifugation (3500 g for 8 min). Repeat this step once more.

Resuspend the pellet after the last step in 1–5 mL cold 1:1 PBS/EtOH and mix.

The fixed sample can be kept in the freezer (−20°C) for several months.

7.3.2 Sample preparation

The sample can typically be applied either directly to the slide to maintain spatial resolution within the flocs, or else after gentle homogenization to get a higher resolution and thus more accurate enumerations (see also Chapter 8). Gentle homogenization can be obtained by gently rubbing two glass slides with 20 μL sample against each other, or more efficiently by a tissue grinder.

Cryosectioning can be carried out to maintain high resolution and still keep the spatial resolution. Cryosectioning can be obtained by embedding the sample in paraffin or in cold polymerizing resin (Tissue-Tek O.C.T.). Paraffin embedding usually provides accurate and homogenous slices without breakage of the biomass sample, but requires heating the sample and de-waxing with xylene, while cold polymerizing resin application can be carried out at low temperature and does not require further chemical treatment.

The biofilm/sludge samples (e.g. in the lid of an Eppendorf tube) are covered by embedding material (Tissue-Tek O.C.T.) which is allowed to migrate into the sample overnight and polymerize at 4°C, before transferring it to liquid nitrogen. Sectioning into 5–20 μm thin slices is carried out on a cryotome at −20°C (e.g. Dublier *et al.*, 1995; Moter and Gobel, 2000; Gieseke *et al.*, 2005). The slices are immediately placed on a slide (room temperature), where they melt. The slices are allowed to dry on the bench for 3 hours.

7.3.3 Immobilization of the samples on glass slides

To provide the best possible adhesion without loss of sample during the further treatments and to obtain an even and homogeneous distribution of the sample, it is recommended to use pretreated glass slides. The best results are usually obtained on a relatively hydrophilic surface, which can be obtained by acid-washed slides and slides coated with gelatin or poly-L-lysine. An acid wash before coating is recommended.

Acid wash procedure

Place the slides in a beaker containing preheated (60°C) 1 M HCl solution.

Let the slides stand for at least 8 hours. After cooling the slides are carefully rinsed in dH_2O.

Rinse the slides in 95% ethanol and let dry.

Procedure for gelatine coating of microscopic slides

Rinse the slides (either normal glass slides or Teflon coated slides) in 70% ethanol, let air dry, dip in 0.5% gelatine solution for 5 min at 70°C.

Remove the slides and let them air dry in vertical position in a dust-free environment.

Procedure for Poly-L-Lysine coating of microscopic slides

The acid-washed slides are incubated in 0.01% poly-L-lysine solution for 5 min at room temperature.

Remove the slides and let them air dry in vertical position in a dust-free environment.

Immobilization of the sample on slides
Place 5–15 μL of the sample on a slide, cover glass or in each well in a teflon-coated slide.
Use the side of the pipette tip to spread out the sample. Let the samples dry in the fume hood until they are completely dry (15–30 min). Drying at elevated temperature (40–60°C) improves the binding of dilute samples to gelatine-coated slides.

7.3.4 Dehydration

Dehydration of the immobilized (untreated, homogenized or cryosectioned) sample removes water from the sample in order to increase the resolution during microscopy. Serial treatment with increasing ethanol concentrations efficiently removes the water in the sample rendering the final sample with a decreased thickness down to almost 25%. Dehydrate the slides with the immobilized sludge samples by dipping it into a container with 50% ethanol for 3 min, followed by 3 min in 80% ethanol and by 3 min in 96% ethanol. Let the slides air-dry before continuing with the permeabilization step.

7.3.5 Permeabilization by enzymatic or chemical treatment

Some cell types require additional treatment by enzymes or chemicals for sufficient permeabilization over the cell wall. Several enzymes have been applied and must be individually tested for new cell types, although a few guidelines exist. Lysozyme treatment makes many Gram positive cells permeable, some require treatment with acid, lipase or mutanolysin (Davenport *et al.*, 2000). Some organisms (e.g. certain *Archaea*) are not readily amendable to the lysozyme permeabilization strategy and require protease treatment, e.g. protease K, mutanolysin or achromopeptidase, while others (e.g. Mycolata) require a mixture of the various enzymatic treatments and/or mild acid hydrolysis pre-treatment (e.g. Carr *et al.*, 2005, Kragelund *et al.*, 2007b). Digestion with enzymes, too long incubations or treatment with too harsh chemicals is critical and will cause lysis of some more sensitive cells. It might thus be incompatible to access multiple organisms in complex environments. Therefore always the gentlest treatment possible should be used.

Procedure for permeabilization with Lysozyme
Start with immobilized samples dehydrated in EtOH series.
Apply 10–15 μL cold lysozyme (36 000–360 000 U/mL) per slide (or well on slide) and place the slide in a horizontal position, in a 50 mL polyethylene tube, containing tissue paper with 2 mL of dH_2O.
Incubate the slides for 10–60 min at 37°C.
Wash the slide three times in dH_2O, then once in absolute ethanol, and let the slide air-dry.
After permeabilization, the slide can be stored at –20°C for several months.

Permeabilization with Mutanolysin
Apply 10–15 μL cold mutanolysin solution (5000 U/mL) in per slide (or well on a slide) and place the slide in a horizontal position, in a 50 mL polyethylene tube, containing tissue paper with 2 mL of dH_2O.
Incubate the slides for 10–30 min at room temperature.
Wash the slide three times in dH_2O, then once in absolute ethanol, and let the slide air-dry.
After permeabilization, the slide can be stored at –20°C for several months.

Procedure for permeabilization with Proteinase K
Start with immobilized samples dehydrated in EtOH series.

Apply 10–15 μL cold proteinase K (2000–20 000 U/mL) per slide (or well on slide) and transfer the slide to a 50 mL polyethylene tube with a moisturized piece of tissue paper. Incubate the slide for 20–60 min at 37°C.

Wash the slide three times in dH_2O, then once in absolute ethanol, and let the slide air-dry.

After permeabilization, the slide can be stored at –20°C for several months.

Permeabilization with Achromopeptidase

Apply 10–15 μL cold achromopeptidase solution (60 U/mL) per slide (or well on slide) and place the slide in a horizontal position, in a 50 mL polyethylene tube, containing a piece of tissue paper with 2 mL of dH_2O.

Incubate the slides for 20–60 min at 37°C.

Wash the slide three times in dH_2O, then once in absolute ethanol, and let the slide air-dry.

After permeabilization, the slide can be stored at –20°C for several months.

Permeabilization with Lipase

Apply 10 μl (75 U/μL) per slide (or well on a slide) and place the slide in a humid chamber containing adsorbent paper wetted with PBS to stop evaporation of the enzyme solution.

Incubate the slides for 60 min at 37°C.

The lipase solution is removed after incubation by washing the slide in dH_2O. Then wash the slide once in absolute ethanol, and let the slide air-dry.

After permeabilization, the slide can be stored at –20°C for several months.

Permeabilization with mild acid hydrolysis

The microscopic slides spotted with dehydrated cells are subjected to hydrochloric acid (1M HCl) at 37°C for 30 min.

Wash the slide with dH_2O, then once in absolute ethanol, and let the slide air-dry.

7.3.6 Preparation and quality check of probes

Commercially acquired probes typically arrive lyophilized and must be resuspended in TE buffer. Stock solutions are diluted to contain 5 μg/μL. Working solutions are prepared at a concentration of 50 ng/μL (by diluting stock solution 1:100 in dH_2O) and stored in small portions (50 to 100 μL) at −20°C.

To test if the probe concentration corresponds to the concentration claimed by the manufacturer the probe is suspended in 50 μL of dH_2O. The absorbance of the 1:100 diluted stock solutions (in dH_2O) are then measured at 260 nm (1 $A_{260} \approx 20$ μg/mL of single-stranded DNA oligonucleotide).

To check the quality of the labeling of the oligonucleotide, the ratio of absorption of fluorochrome and DNA (presuming optimal labeling), i.e. A_{260}/A_{550}, should amount to ≈ 1 for a Cy3-labeled 18-mer probe.

7.3.7 Hybridization

Hybridization of the oligonucleotide probe to the ribosome in the target cell must be carried out under high stringency, with defined conditions regarding preheating temperature, ionic strength, and proper formamide concentrations. By providing the proper conditions, sufficient stringency can be obtained to discriminate between single mismatches in the hybridized probe-ribosome complex.

Prepare 2 mL of hybridization buffer for each percentage of formamide (see Table 7.1). The formamide concentration must reflect the empirically optimized conditions for the applied gene probes. For details on many published probes see chapters elsewhere in this book or visit probeBase (http://www.probeBase.net; Loy *et al.*, 2007).

Table 7.1. Composition of the hybridization buffer at different formamide concentrations.

FA (%)	FA (μL)	Milli-Q (μL)	5 M NaCl (μL)	1 M Tris/HCl (μL)	10% SDS (μL)
0	0	1600	360	40	2
5	100	1500	360	40	2
10	200	1400	360	40	2
15	300	1300	360	40	2
20	400	1200	360	40	2
25	500	1100	360	40	2
30	600	1000	360	40	2
35	700	900	360	40	2
40	800	800	360	40	2
45	900	700	360	40	2
50	1000	600	360	40	2
60	1200	400	360	40	2
65	1300	300	360	40	2
70	1400	200	360	40	2

Hybridization procedures are typically designed to occur at 46°C, but can be carried out at other temperatures by changing the matching formamide concentration according to the relation: 1% FA = 0.65°C.

Transfer 8 μL of hybridization buffer onto the slide within a small area or into each well on a teflon-coated slide. Prepare one slide at a time to avoid evaporation of hybridization buffer and thus changed stringency.

Add 1 μL of each gene probe (work solution 50 ng/μL) and mix carefully (avoid contact with the sample) with the hybridization buffer (sterile pipette tips must be used for all work with gene probes). If more gene probes are added to the same well the order is in the exact same manner – 1 μL of each of the probes is added to the well. Equimolar concentrations of each competitor probe are added if needed.

Place the slide horizontally into a 50 mL polyethylene tube with a piece of tissue wetted with 1–2 mL of hybridization buffer. Place the tube in the hybridization oven (46°C) for at least 1½ hours (increased hybridization to less accessible regions of the ribosome has been shown to occur upon hybridization for up to 72 hours, see also Chapter 5).

Probes with different T_d (requiring different formamide concentrations) cannot be applied together, but must be applied in a double hybridization with two subsequent hybridizations starting with the highest formamide concentration.

During hybridization prepare the washing buffer in a 50 mL polyethylene tube (formamide is replaced by NaCl, according to Table 7.2). Preheat the washing buffer in a 48°C water bath.

Gently rinse the slide by pouring a few millilitres of the preheated washing buffer over the sample (must be carried out in the fume hood to avoid toxic formamide fumes).

Transfer the slide to the 50 mL polyethylene tube with the remaining preheated washing buffer, and incubate for 15 min at 48°C (water bath).

Remove the slides from the washing buffer and dip in cold dH_2O. Let the slide air-dry.

Table 7.2. Composition of the washing buffer corresponding to the formamide concentrations applied during the hybridization.

FA (%)	1 M Tris/HCl pH 8.0 (μL)	10% SDS (μL)	5 M NaCl (μL)	0.5 M EDTA (μL)
0	1000	50	9000	0
5	1000	50	6300	0
10	1000	50	4500	0
15	1000	50	3180	0
20	1000	50	2150	500
25	1000	50	1490	500
30	1000	50	1020	500
35	1000	50	700	500
40	1000	50	460	500
45	1000	50	300	500
50	1000	50	180	500
55	1000	50	100	500

7.3.8 Counterstaining with DAPI

After hybridization the slide can be counterstained, for example, with the DNA stain 4′,6′-diamidino-2-phenylindol (DAPI) to determine the fraction of FISH positive out of the total DAPI count (optional).

Add DAPI solution (1 μg/mL in distilled water) to cover the sample, and stain for 15 min at 4°C in the dark.

Rinse with plenty of dH_2O, and let the slide air-dry.

Please note that DAPI is staining DNA only, for other counterstaining methods see Chapter 8.

7.4 MICROSCOPY

Evaluation of the FISH signal is performed after embedding in a small drop of mounting agent containing an antibleaching agent (e.g. Citifluor or Vectashield). Volumetric mixtures of Citifluor and Vectashield are preferred by several users (optional).

The embedding medium can also be directly amended with DAPI (final concentration 1 μg/mL). This usually yields strong signals but renders a strong background fluorescence. Stained preparations can also be stored at –20°C without substantial loss of signal intensity.

Upon examination of multiple stained slides always examine the longer wavelength first (less energetic) to avoid bleaching.

When examining new samples always include a nonsense probe (Non-EUB338) using the same strong fluorescing fluorochrome as the one used with the specific probes (e.g. Cy3 or Alexa-dyes). A detailed review of the factors influencing the sensitivity of FISH, effect of fluorochrome type, and stringency conditions are described by Bouvier and del Giorgio (2003).

7.5 RECOMMENDATIONS AND TROUBLESHOOTING

Here follows a list of typical problems encountered when working with the FISH technique such as low signals and problems related to the microscopic examination.

Test of new oligonucleotide probes

Several software programs exist for design of new probes (e.g. ARB software package (http://www.arb-home.de), and Primrose (Ashelford *et al.*, 2002)). Several publications describe the choice of parameters and pitfalls during the process of designing new probes (e.g. Amann and Ludwig, 2000) and only a few recommendations for the testing will be given here.

- New probes should always be empirically tested for optimizing the applied stringency (i.e. the formamide concentration under which it should be applied). This test is performed by measurement of fluorescence signal intensities after FISH with increments of 5% formamide on a pure culture with zero, one and two mismatches (harvested by the end of the exponential growth phase). The fluorescence intensities of individual cells on images acquired under identical settings are quantified by image analysis, and plotted as function of the formamide concentration (see Chapter 8). The adequate hybridization stringency is the highest concentration of formamide that does not result in a loss of signal intensity of the target cells.
- In cases where no pure culture exists with a perfect match and with one mismatch to the probe target sequence, as often seen in the cases where the probe was designed from clone libraries, validation and optimization can be performed by Clone-FISH (for an example see Kong *et al.*, 2005).
- Alternatively, the probe can be applied in complex microbial environments, such as activated sludge. Fluorescence signal intensities and cell numbers after FISH with increments of 5% formamide are measured by image analysis. Cells with mismatches are targeted (due to too low stringency) when the numbers of cells with a certain threshold of fluorescence begin to increase. The proper stringency is found at the formamide concentration when the number of FISH positive cells starts to decline.
- *In silico* predictions of the proper hybridization stringency are still not sufficiently accurate, and thus optimization of probe specificity and sensitivity should be performed experimentally.
- Mismatch discrimination can be based on differences in formamide denaturation profiles or by use of competitor probes (recommended when $\Delta FA < 20\%$).
- Check the ability of a new probe to form hairpin structures or self annealing. This can be checked by appropriate software such as DINAMelt (Markham and Zuker, 2008).

Microscopy

- Loss of cells during hybridization can be minimized by acid wash and gelatin coating of the slides. Gentle heating (40–60°C) during immobilization also decreases cell loss. If this is not sufficient then a low-temperature-melting agarose on top of the immobilized cells can reduce the loss of cells. Agarose embedding will give less sharp images, although still usable for quantification purposes (see Chapter 8).
- Blurred images are often seen when adding too much or too little antifading agents and/or immersion oil.
- Keep the illumination of the excitation light to a minimum, that is, block the light path when not directly examining the sample.
- Always check for autofluorescent cells or debris by checking for fluorescence at other wavelengths than the one used for the probe. Autofluorescence typically (but not always) has a broad emission spectrum.
- In case of too high autofluorescence, choose narrow bandpass filters instead of longpass filters.

- Too high background can be avoided by using less or thinner biomass samples. Longer washing times can sometimes be helpful. Large flocs can be homogenized by robbing two glass slides with 20–40 μL of the sample towards each other (circular movements). Many automated sample homogenizers are seldom optimal for sludge samples (either too harsh or too gentle). In our lab we have successfully applied a tissue grinder, which can be obtained in different sizes and volumes.
- Some fluorochromes simply do not work with certain samples, either because of precipitation or autofluorescence.

Low signals or no signals

- Inaccessibility of the probe can cause absent signals or low fluorescence. This can be overcome by application of helper probes (Fuchs *et al.*, 2000). Helper probes are unlabeled probes designed to target rRNA sequences adjacent to the FISH probe target site, and are believed to open secondary and tertiary structures and thereby increase accessibility of the FISH probe. Another solution to inaccessible sites has been shown to be a longer hybridization time (Yilmaz *et al.*, 2006). Increasing the hybridization time (for up to 72 hours) will improve probe diffusion into the cell and decrease kinetic barriers of target site accessibility. Thus a better hybridization efficiency and in general increased fluorescence signals are achieved.
- Insufficient cell wall permeability described for the Mycolata group, for example, can also limit target accessibility causing no or low FISH signals. The application of different pre-treatment protocols as described in this chapter can improve cell wall permeability for FISH.
- The choice of fluorochrome is important if low fluorescent signals occur. The larger the extinction coefficients are, the stronger signals can be obtained. In Table 7.3 the most often used fluorochromes are listed together with their extinction coefficients.
- Avoid too long exposure to strong light sources. Store the FISH slides in the dark.
- Carefully remove all traces of ethanol from stored samples prior to DAPI staining or FISH, as ethanol can fade the fluorescence signal.
- Avoid too long fixation time, which (especially for PFA fixation) can reduce permeability.

Table 7.3. Most commonly used fluorescent dyes to label oligonucleotides for FISH analysis. For optimal filters to visualize the above given dyes see the webpages of the main suppliers (e.g. http://www.zeiss.de; http://www.chroma.com).

Fluorochrome	Colour	Max Excitation λ (nm)	Max Emission λ (nm)	Extinction coefficient ($M^{-1}cm^{-1}$)
Cy3	Red	552	565	150,000
Cy5	Red[a]	649	670	250,000
SYBR Green	Green	494	520	73,000
DAPI	Blue	350	456	27,000
FLUOS	Green	494	523	74,000
TAMRA	Red	543	575	65,000
Alexa-488	Green	493	517	71,000
Alexa-546	Red	562	573	104,000
Alexa-350	Blue	343	441	19,000

[a] emission in the infrared area, it thus requires a digital camera detecting infrared light and image analysis to be observable

General planning

- Always test your sample for unspecific binding by applying a nonsense target probe (e.g. NonEUB; Stahl and Amann, 1991). Use the same fluorochrome as used for the specific probes to be analyzed.
- Although many Gram positive cells can be seen using a protocol for fixation of Gram negative cells (and vice versa), quite a few might be impaired due to clogging of the cell membrane or cell lysis. No universal fixation exists, so use the proper fixation procedure for the cells of interest. For quantification purposes, stick to the recommendations in Chapter 8.
- Use (if possible) multiple probes targeting different phylogenetic levels, in order to verify that the specificity of the most specific probe (e.g. all genus/strain-specific probes) should also be targeted by the phylum/order/family-targeted probes.
- Hybridization efficiency depends on the hybridization length and should be optimized on new samples.
- Specificity and coverage of new and old probes should be tested on a regular basis. This can be done using various freely accessible internet tools [e.g. probeBase (http://www.probeBase.net) or Probematch (http://rdp.cme.msu.edu/)].

8

Quantitative FISH for the cultivation-independent quantification of microbes in wastewater treatment plants

Holger Daims

8.1 INTRODUCTION

In many fields of microbial ecology and environmental engineering, including research related to wastewater treatment, not only the cultivation-independent detection but also the *in situ* quantification of uncultured microorganisms are important tasks and the main reasons to use molecular biological methods. For example, monitoring the abundances and population dynamics of selected microbes in activated sludge can give important clues on correlations between problems such as sludge bulking and the microbial community composition. As outlined in the other chapters of this book, fluorescence *in situ* hybridization (FISH) with rRNA-targeted probes is a very powerful tool and the method of choice for the cultivation-independent detection and identification of microorganisms in activated sludge and biofilm from wastewater treatment plants (WWTPs).

A key advantage of FISH over other methods, for example those based on PCR, is the direct visual feedback one gets during microscopic observation of the analyzed samples. This does not only give more confidence regarding the accuracy of an experiment, but it is also the basis for extensions that convert FISH into a precise method to quantify microorganisms in environmental samples. Based on the rule "count what you see", quantitative FISH often is less prone to biases than quantitative PCR approaches, which are quite sensitive to methodical errors (e.g. low PCR efficiency or stochastic tube-tube variations) and laboratory contaminations with tiny amounts of template DNA. Furthermore, PCR-based approaches do not actually quantify microbial cells, but measure copy numbers of marker genes. The number of marker gene copies per genome, and even the number of genomes per cell, can vary among the targeted microbes (Ludwig and Schleifer, 2000). Thus, measuring these copy numbers by PCR allows only indirect conclusions on cell densities and on important parameters such as substrate turnover rates per cell. In contrast, quantification methods based on FISH directly measure cell numbers or equivalent quantities such as biovolume. A disadvantage of FISH is its lower sensitivity compared to quantitative PCR: while PCR can (theoretically) detect a single microbial cell, the detection limit of FISH is as high as 10^3–10^4 cells per ml of suspended activated sludge. On the other hand, less frequent populations are unlikely to be functionally relevant for the wastewater treatment process. Cultivation-based methods, for example most-probable number (MPN) counting, also are potentially more sensitive than FISH but do not correctly quantify aggregated cells and fail to detect all organisms that do not grow in the offered nutrient media. Thus, FISH is the best currently available technique for quantifying microbes in lab-scale or pilot-scale reactors and in samples from full-scale WWTPs.

This chapter briefly summarizes approaches to quantify microbes by FISH with special focus on combining FISH with digital image analysis. A protocol is provided that uses a slightly modified FISH protocol and free open-source software for measuring the abundances of probe-target populations in activated sludge and biofilm samples.

8.2 QUANTITATIVE FISH: A BRIEF OVERVIEW

The simplest approach to quantitative FISH is direct visual counting of probe-stained microbial cells under the fluorescence microscope. For this purpose, FISH is performed according to the standard protocol as described elsewhere in this book. Subsequently, the probe-target cells are counted in several microscope fields of view (FOV). Unfortunately, the obtained numbers can hardly be related to any useful reference value, such as sludge volume, if complex samples like activated sludge or biofilm are analyzed this way. This is due to (i) difficulties of applying an exact known amount of flocculated biomass onto a glass slide prior to FISH, and (ii) variations in biomass content between different FOV. Regarding point (i), please note that by FISH only a few microlitres of sludge are analyzed in each experiment. Even though standard methods allow the determination of biomass concentration for larger volumes, variations due to pipetting errors and differences in the sizes of individual flocs become a major source of imprecision when very small volumes are handled. Therefore, most past studies determined relative rather than absolute abundances of microbes in activated sludge. In this case, the sample is hybridized to a specific probe and also to a ‘universal’ second probe or probe mixture, which targets all *Bacteria*, like the EUB-mix (Daims *et al.*, 1999). Alternatively, the ‘universal’ probe(s) can be replaced by a nucleic-acid staining dye such as DAPI. Following FISH, the cells stained by the specific probe and the cells stained by the ‘universal’ probe or dye are counted. This is done for several FOVs and the results are summed up to determine the fraction of the specifically stained cells relative to all microbes at the end. This technique has been used, for example, to quantify non-aggregated cells in activated sludge (Wagner *et al.*, 1994b). While simplicity is its greatest advantage, this method is very time-consuming and tedious, because several hundreds or even thousands of cells must manually be counted in several FOVs in order to obtain statistically robust results. It is not suitable to quantify cells in aggregates, because a precise counting of clustered cells, by eye and in many FOVs, is hardly feasible. The disruption of tight cell aggregates in activated sludge to facilitate counting is not possible either (Manz *et al.*, 1994; Wallner *et al.*, 1995).

The spread of digital imaging techniques and equipment has paved the way for combining rRNA-targeted FISH with automated image analysis methods. For example, Pernthaler *et al.* (2003) designed a fully automated setup that linked image acquisition by microscopy to an image analysis pipeline for counting planktonic bacterial cells in ocean water samples. This sophisticated approach was much more rapid than manual counting and enabled the analysis of 28 samples within only 4 hours. Unfortunately, microbes that cluster in biofilms cannot precisely be counted by this and related methods. Prior to cell counting by image analysis, the images must be *segmented* by the computer – that means, the machine must detect the objects to be counted in the images. In case of planktonic cells, which do not form tight aggregates, every detected object corresponds to one single cell (or very few attached cells). In contrast, the individual cells embedded in tight cell clusters cannot be distinguished by image segmentation in most cases, because in the images there is no background that would separate these cells from each other. Thus, a fully automated quantification of cells in sludge flocs or biofilms usually is not possible.

An alternative solution is to measure the biovolume fraction of the targeted microbial population relative to the total microbial community. This approach was developed and first applied on activated sludge by Bouchez *et al.* (2000) and Schmid *et al.* (2000). The sludge sample is hybridized to a specific probe and a 'universal' probe (set) as described above for manual counting. Subsequently, digital images of either probe signal are recorded for 20–30 FOVs. This is best accomplished by using a confocal laser scanning microscope (CLSM) to avoid blurred images due to light from outside the focal plane. However, a normal epifluorescence microscope with a CCD camera can be used if low amounts of biomass are applied onto the slides to obtain thin layers of biomass, which are completely within the depth of focus of the microscope. An experienced user can record the images needed to quantify one population in about one hour. Importantly, the FOVs must be randomly selected in the x, y and z dimensions (by panning and zooming with the microscope). When all images are taken, the areas of the probe-target organisms are measured, by image analysis, in the images showing the signal of the specific probe and are summed up to obtain a total measured area of the probe-target population. Likewise, the areas of all organisms detected by the 'universal' probe are measured in the respective images. The area fraction of the specifically targeted population is then calculated as percentage of the total area of all organisms. Although these measurements are made with two-dimensional images (not with confocal z-stacks), the determined area fraction is an estimate of the biovolume fraction of the quantified population (Daims and Wagner, 2007). Notably, this approach does not require the detection of individual cells during image segmentation. Areas in images can be measured for single cells as well as for filaments or cell aggregates. This greatly simplifies image segmentation and allows even densely packed biofilm bacteria to be quantified.

The primary result of this 'biovolume-based' quantification method is a relative quantity, the biovolume of the probe-target population as percentage of the total microbial biovolume in the sample. This is sufficient for many applications such as monitoring population dynamics. However, absolute cell numbers are needed to estimate substrate turnover rates per cell, and also to compare different sludge samples with each other (because the total biomass content of these samples may be different). Absolute cell numbers can be obtained by an extended 'biovolume-based' approach called 'Spike-FISH', which uses an internal standard to convert biovolume fractions to absolute cell numbers (Daims *et al.*, 2001b).

A more detailed description of the quantification methods mentioned here, and of additional techniques, is provided by Daims and Wagner (2007). The next section explains how the 'biovolume-based' quantification method is used in practice to determine relative amounts of probe-defined bacterial populations in activated sludge.

8.3 A PROTOCOL FOR QUANTITATIVE FISH AND IMAGE ANALYSIS TO MEASURE BIOVOLUME FRACTIONS

Sample preparation

For quantitative FISH follow the general sample fixation procedures described in this book in Chapter 7. Gram negative and Gram positive cells should be quantified in separate steps using formaldehyde-fixed or ethanol-fixed sample aliquots, respectively. Formaldehyde fixation usually applied to preserve Gram negative cells renders most Gram positive bacteria impermeable to FISH probes. In contrast, ethanol fixation used for Gram positive cells does not preserve Gram negative cells for extended time periods and when these cells start to lyse, the quantification results will most likely become biased. Moreover, if additional enzymatic pre-treatment is needed to permeabilize Gram positive cells, the Gram negative microbes are easily destroyed (lysed) by the enzymes. In cases where only ethanol-fixed samples are available, these samples should be analyzed within **1–3 days** after sampling and fixation to reduce biases due to lysed Gram negative cells.

Sample immobilization

For standard FISH only small amounts (5–15 μl) of an activated sludge sample are immobilized on glass slides. This is sufficient for purely qualitative analyses, but can cause biases in quantifications due to single cells falling down onto the slide surface (Figure 8.1A). To avoid this problem, thick layers of sludge sample must be applied onto the slides (Figure 8.1B). First, put 15 μl of fixed sludge onto the slide and dry the sludge at 46°C or at room temperature. Subsequently, add another 15 μl aliquot and dry again. These steps should be repeated two to four times to obtain a thick layer of biomass (Figure 8.1B). With this large amount of sludge on the glass slide, additional measures are needed to avoid a loss of biomass during the hybridization and washing steps. First, poly-L-lysine-coated slides should be used to improve sludge adhesion to the slide surface. Second, the dried biomass should be covered with a thin agarose layer, which will act like a glue and keep the biomass attached to the slide (Figure 8.1C). For this purpose, dip the slide (with the dried biomass) horizontally into hand-warm dissolved 0.5–1% agarose, leave it for a few seconds in the agarose, and then cool the slide immediately in the horizontal position (biomass side up) on ice. When the agarose has solidified, the usual dehydration in 50, 80 and 96% ethanol is performed and the slide is dried again at 46°C. The combination of dehydration and drying will shrink the agarose so that a thin, glue-like agarose film is obtained on top of the sludge sample (Figure 8.1C). Finally, wipe off excess agarose from the bottom side of the slide. Although agarose may show autofluorescence, the very thin agarose layer does usually not interfere with fluorescence microscopy.

In situ hybridization

Hybridization, washing, drying, and embedding the agarose-covered slide in antifadent are carried out according to the standard FISH protocol. For quantitative FISH, one must apply (i) a specific probe that targets the population(s) to be quantified, and (ii) a 'universal' probe set that detects most *Bacteria*, which usually is the EUB-mix (Daims *et al.*, 1999). Probes (i) and (ii) must be labeled with different fluorochromes. The 'universal' probe can be replaced by a nucleic acid stain, but be sure to use only dyes that stain both RNA and DNA. For example, DAPI is not useful for this kind of quantification, because DAPI stains DNA only. Unlike ribosomes, DNA is not evenly distributed inside most bacterial cells. Therefore, DAPI-stained cells look smaller than cells labeled by FISH and their projected area in the recorded images is smaller. Clearly, this biases a quantification technique based on measuring cell areas in the images. In contrast, SYBR Green stains RNA as well as DNA and can be used for this quantitative FISH approach.

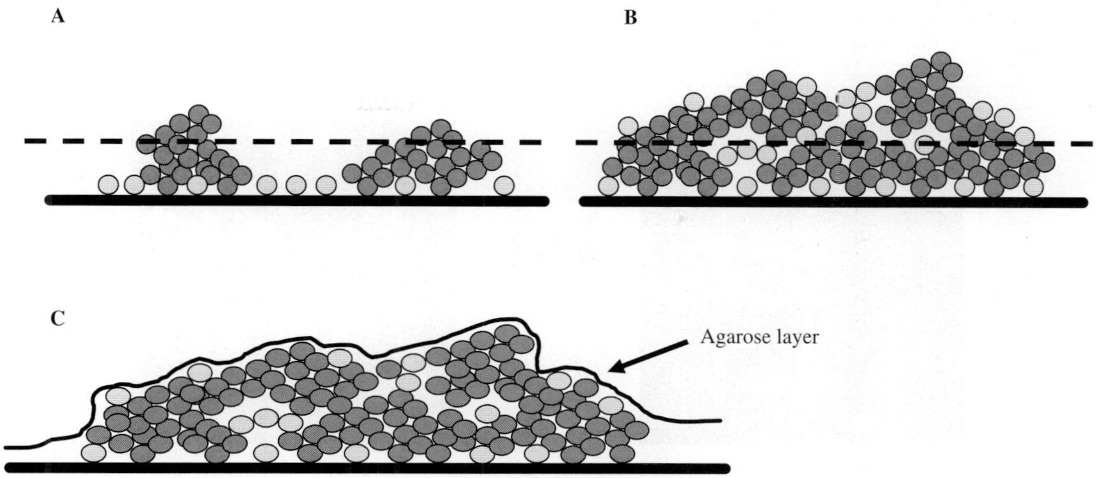

Figure 8.1. A. Low amount of activated sludge biomass on a glass slide. During dehydration for FISH, single cells (light grey) may fall down onto the glass slide, whereas cell aggregates (dark grey) stay in place. Depending on the focal plane of the microscope (stippled line), mainly aggregates or single cells are captured in the images. **B.** More biomass on the slide prevents most small cells from falling down onto the glass surface. **C.** Covering the biomass with a thin agarose layer as 'glue' prevents biomass detachment from the slide during the dehydration, hybridization, and washing steps.

Image acquisition

Images for quantitative FISH should be recorded by using a CLSM. Take images of 20–30 FOVs, which must be chosen by randomly moving the slide in the *x*/*y* dimensions and by randomly adjusting the focal plane of the microscope in the *z* dimension. 'Randomly' means: do **not** choose the FOVs based on the presence of the population to be quantified! This would bias the analysis. Make sure that each FOV contains some microbial biomass (not necessarily the quantified population), but do not use any other selection criteria.

For each FOV, take one image of the population-specific probe and one image of the 'universal' probe (or nucleic acid stain). Use a low magnification (400× or lower) to capture as much biomass as possible in each image. This will improve statistics. Cells of the quantified population will appear in either image of an 'image pair' (specific probe and EUB-mix; Figure 8.2) in different colors. The detector of the CLSM must be adjusted properly to ensure that these cells have the same size (area) in both images. If this condition is not met, the analysis will most likely be biased. Please keep in mind that this quantification method is based on measuring the areas of cells in the images and on calculating area fractions (see above).

It makes sense to store the image pairs with numbered filenames to facilitate later analysis. For example, the two images Gamma_001.tif and EUB_001.tif would belong to FOV #1 (with *Gammaproteobacteria* as the organisms to be quantified).

Image analysis

Measuring the areas of the probe-stained cells essentially means counting and summing up the pixels that belong to these cells in each image. First, the images must be segmented (see also below) to distinguish cells (= 'objects') from background. Second, the pixels that belong to objects must be counted.

Commercial image analysis programs, which are capable of automated image segmentation and calculations such as area measurements, are mostly very expensive. But for quantitative FISH there is no need to invest in such software, because some free software packages are available that can fulfil these tasks as well. Examples are the programs CMEIAS (Liu *et al.*, 2001a), ImageJ (http://rsbweb.nih.gov/ij/),

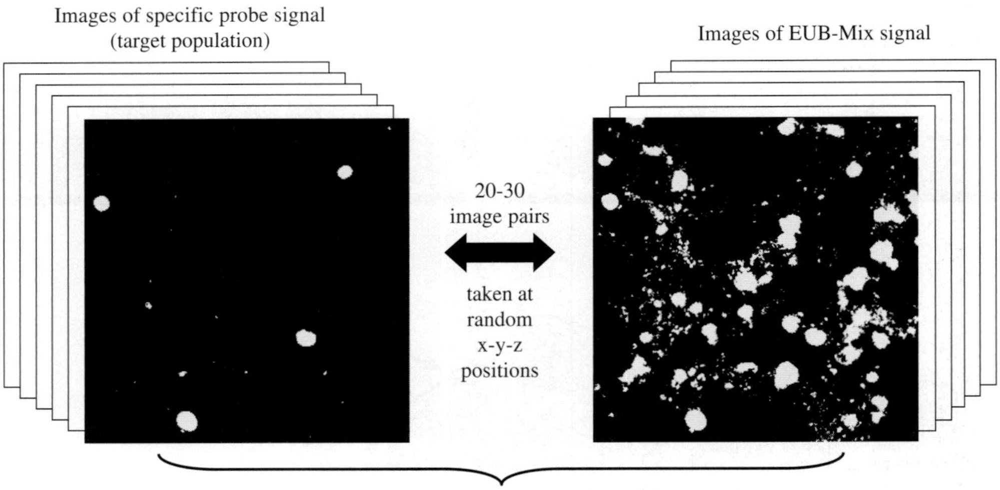

Area of probe-target population (A^P) and of all bacteria (A^B) measured in n pairs

$$\text{Biovolume fraction} \approx \frac{\sum_{i=1}^{n} A_i^P}{\sum_{i=1}^{n} A_i^B}$$

Figure 8.2. Principle of approach for quantifying the biovolume fraction of a probe-defined target population. See text for details.

and *daime* (Daims *et al.*, 2006). While ImageJ is a general-purpose image analysis software, CMEIAS and *daime* have been developed having applications in microbiology in mind.

This chapter focuses on using *daime* for quantitative FISH. This program ("**d**igital image **a**nalysis **i**n **m**icrobial **e**cology" = *daime*) has been written especially for analyzing images of microbial cells as obtained by FISH and other fluorescence labeling techniques. It allows images to be segmented by using different intensity thresholding and edge detection algorithms. If needed, background can be reduced prior to segmentation. Segmented images can be opened in an 'object editor', which allows the user to manually refine the automated segmentation results and to remove fluorescent artefacts (such as plant material) from the images before the actual analysis starts (Figure 8.3). *daime* offers various image analysis routines: morphometry of single cells or cell aggregates, biovolume fraction measurement for quantitative FISH, fluorescence intensity measurements, automated analysis of formamide concentration series for FISH probe evaluation, and spatial statistics to determine if two microbial populations co-aggregate in biofilm samples (Daims *et al.*, 2006; Maixner *et al.*, 2006). In addition, *daime* contains a volume-rendering module that can display confocal image stacks in 3D at interactive rendering speed on current PC hardware.

More information about *daime* and its features can be found on the *daime* website (www.microbial-ecology.net/daime) and in the user manual. The program and the manual can be downloaded free of charge. *daime* runs on Windows 2000 and Windows XP (Vista not yet tested) and Linux (the source code is available for compiling *daime* on Linux systems). Appropriate graphics hardware and sufficient RAM memory (at least 1 gigabyte) are recommended for optimal performance.

To analyze the images of the different FOVs, which have been acquired in the previous steps, import them into *daime* as 2D image series. For this purpose, the image files should have been saved to disk with numbered filenames (as suggested above) and in the TIFF image format (8 bits per pixel, greyscale, or

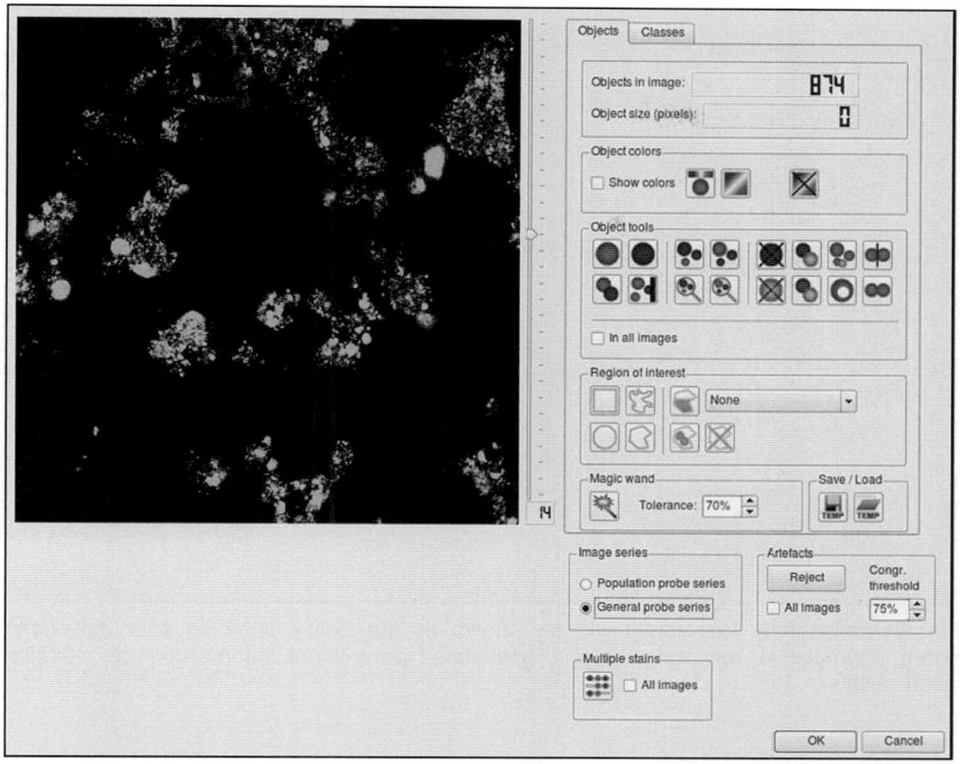

Figure 8.3. Screenshot of the 'object editor' of the *daime* software. This dialog window offers several tools for identifying artefacts in images of probe-labeled microbial cells, for selecting objects based on morphological and other features for subsequent analysis, and for the manual refinement of automated image segmentation.

24 or 32 bits per pixel, RGB color). Then choose the option 'Stereology: Biovolume fraction' in the 'Analysis' menu. Indicate which image series shows the population to be quantified, and which one shows all cells detected by the 'universal' probe. The images will then be segmented automatically, and the object editor will open with both image series loaded (Figure 8.3). Push 'OK' if the segmentation results look satisfying. The areas will now be measured in all images, and the results window will open (Figure 8.4). As described above, the area fraction determined from all images is an estimate of the biovolume fraction of the quantified population. In addition to this value, *daime* provides some other parameters and statistics that allow the reliability of the quantification to be assessed (Figure 8.4). This depends mainly on the quality of the input images and on the success of image segmentation.

8.4 CONCLUDING REMARKS

With today's microscopes and computers, the FISH-based quantification of microbial populations in activated sludge is a straightforward task. However, like any other laboratory technique this approach has pitfalls and suffers from potential biases. Some of these problems have been addressed in the sections above. Here are a few general suggestions on how to optimize quantitative FISH analyses:

- Make sure that the basic method, FISH itself, works properly with your samples. Always test the applied oligonucleotide probes with samples known to contain the target organisms

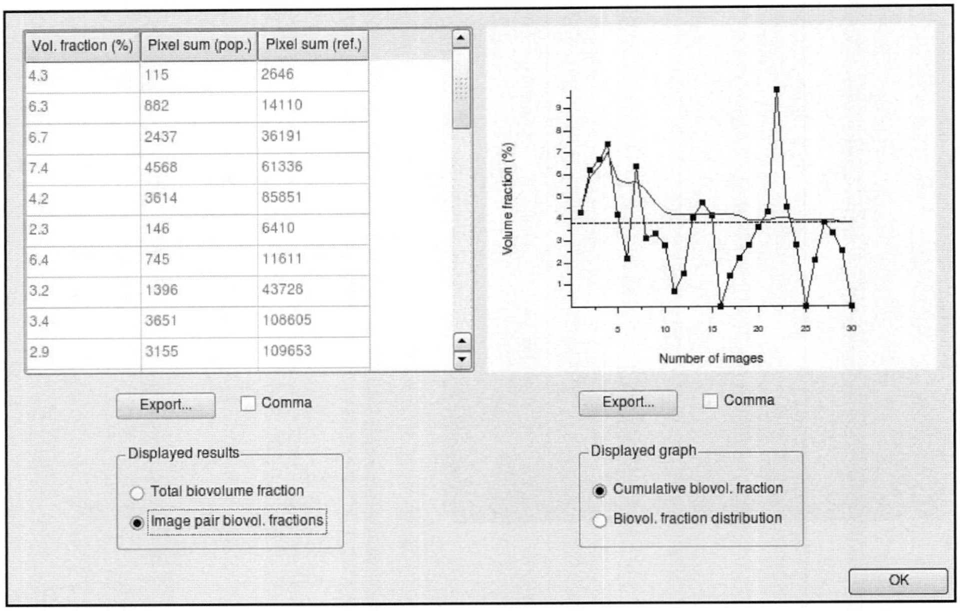

Figure 8.4. Screenshot of the results window shown by the *daime* software after biovolume fraction measurement. Cumulative average analysis and other parameters inform on the reliability of the quantification results.

(positive control). Also use standard negative controls for FISH (e.g. nonsense probes) to test whether autofluorescence or unspecific attachment of probes to cells and to other surfaces might hamper quantification by FISH and image analysis. Typical caveats of FISH and solutions have been addressed elsewhere (Wagner *et al.*, 2003; Daims *et al.*, 2005).

- A low ribosome content of the probe-target organisms is a typical obstacle for the FISH-based quantification of bacteria in oligotrophic samples. As WWTPs are not oligotrophic and most bacteria therein are metabolically active and contain many ribosomes per cell, this problem is hardly encountered when analyzing activated sludge. However, sometimes signal amplification techniques such as CARD-FISH (Pernthaler *et al.*, 2002) are needed to obtain good probe-conferred signals even in samples from WWTPs. The issue of per-cell ribosome content in the context of FISH has been addressed in detail by Hoshino *et al.* (2008).
- Try to achieve the highest possible quality of the digital images that contain the probe-labeled cells. Image analysis software depends on a good quality of the input images. Do not expect marvelous results from images that are dim (low-contrast), blurred, contain intense background, have a low pixel resolution (below 512^2), or contain only very few cells of the target population to be quantified.
- If image analysis does not work as expected, despite a good image quality, remember that there is no perfect algorithm for image segmentation (the step that most frequently causes trouble). Different software packages offer different segmentation algorithms, so take advantage of this diversity and play with the programs and their options. As image analysis is a science in its own right, consulting literature on this topic may be a good idea before implementing an image analysis procedure for routine work in the lab.

9

Color image section

Nitrifiers

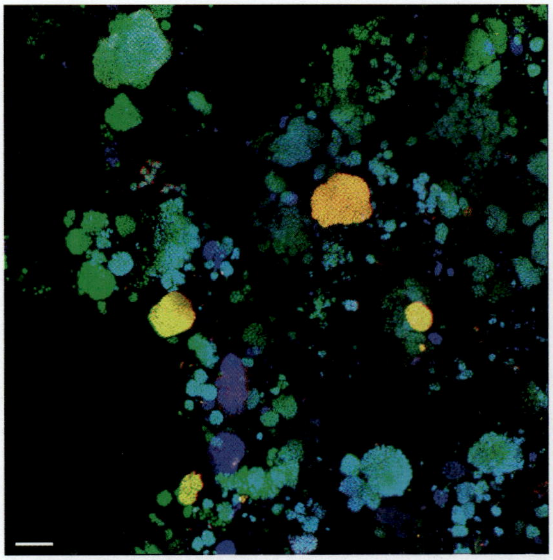

Figure 9.1. Different populations of ammonia-oxidizing bacteria (AOB) simultaneously detected in a nitrifying biofilm from a sequencing batch biofilm reactor with probes NmV (Cy3; red), Nso1225 (Cy5; blue), and NEU (FLUOS; green).

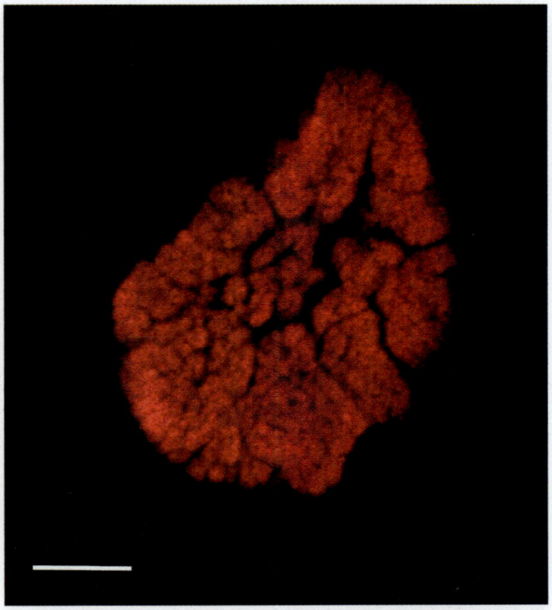

Figure 9.2. Cell aggregate of nitrite-oxidizing *Nitrospira* detected with probe Ntspa662 (Cy3; red) in a nitrifying biofilm from a sequencing batch biofilm reactor. Bar = 10 μm.

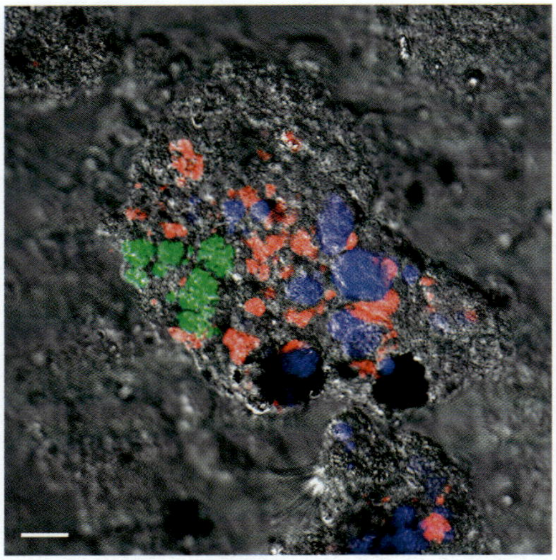

Figure 9.3. Three populations of nitrifying bacteria, detected simultaneously in a nitrifying sequencing batch biofilm reactor with ammonia-oxidizing bacteria (AOB) of the *Nitrosomonas oligotropha* lineage detected by probe Cluster6a192 (Cy5; blue), nitrite-oxidizing bacteria (NOB) of *Nitrospira* sublineage I probe Ntspa1431 (Cy3; red), NOB of *Nitrospira* sublineage II probe Ntspa1151 (FLUOS; green). Bar = 10 μm.

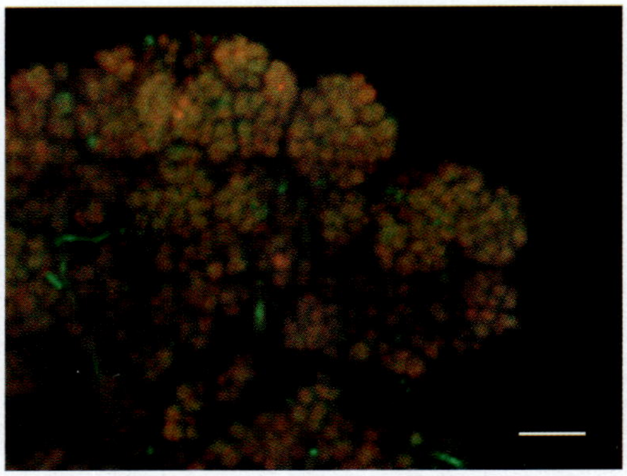

Figure 9.4. Anaerobic ammonium-oxidizing (anammox) bacteria, *Candidatus* Brocadia anammoxidans, detected in an anammox enrichment culture established from a wastewater treatment plant with probes Amx368 (Cy3; red) and EUB338 probe mix (FLUOS; green). Anammox cells appear yellowish due to simultaneous binding of Amx368 and EUB338, whereas other bacteria appear green due to binding of EUB338-I/II/III only. Bar = 5 μm.

Denitrifiers

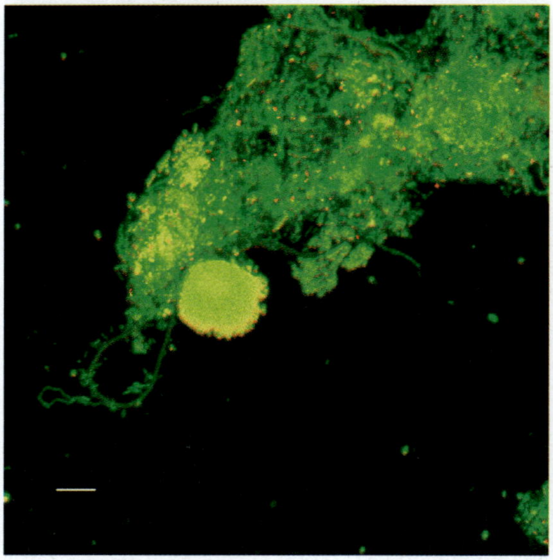

Figure 9.5. Typical morphology of single cells and microcolonies of denitrifying *Curvibacter* spp. detected with probe Curvi997 (Cy3; red) in activated sludge and EUBmix (FLUOS; green). Bar = 10 μm.

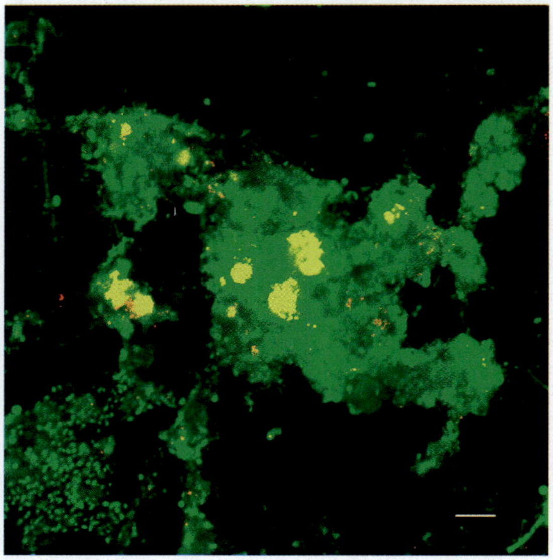

Figure 9.6. Typical cell aggregates of denitrifying *Azoarcus* detected with probe Azo644 (Cy3; red) in activated sludge and EUBmix (FLUOS; green). Bar = 10 μm.

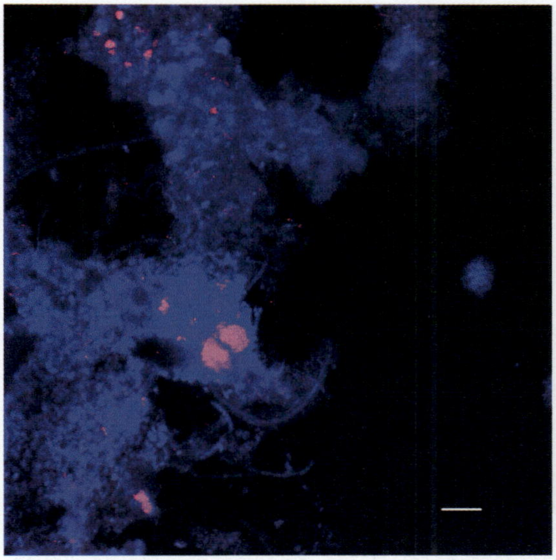

Figure 9.7. Typical cell aggregates of denitrifying *Thauera* spp. detected with probe Thau646 (Cy3; red) in activated sludge and EUBmix (FLUOS; green). Bar = 10 μm.

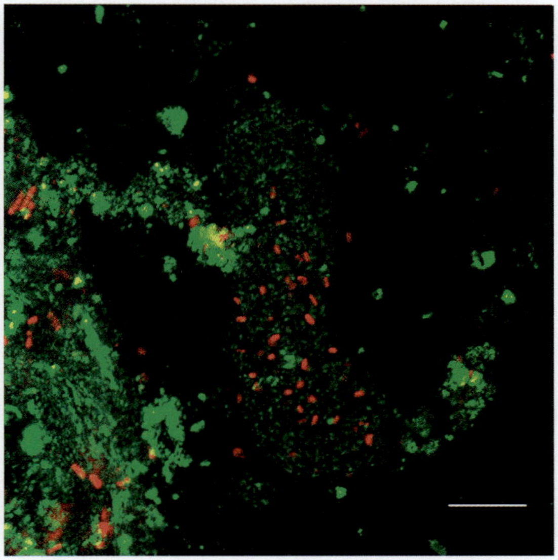

Figure 9.8. Typical morphology of cells belonging to the denitrifying *Acidovorax* spp. detected with probe Aci208 (Cy3; red) in activated sludge and EUBmix (FLUOS; green). Bar = 10 μm.

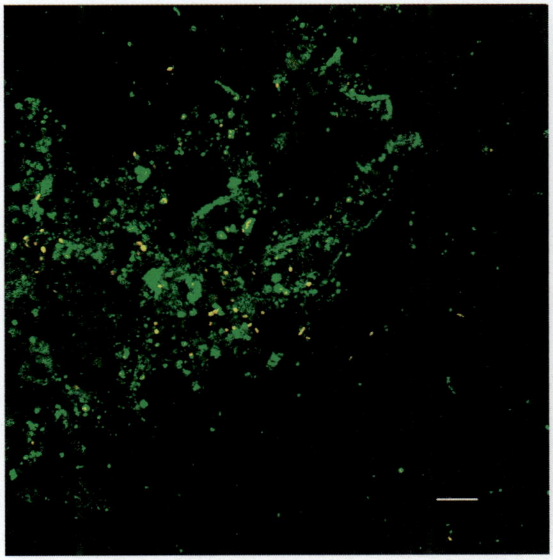

Figure 9.9. Typical morphology of single cells of denitrifying *Hyphomicrobium* spp. detected with probe HyphoCII-654 (Cy3; red) in activated sludge and EUBmix (FLUOS; green). Bar = 10 μm.

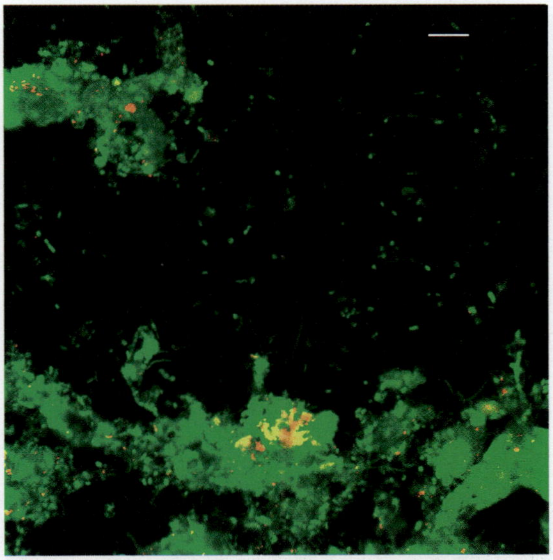

Figure 9.10. Typical cell structures of denitrifying cells belonging to the *Rhodobacteraceae* and *Phyllobacteriaceae* detected with probe GRb 208(Cy3; red) in activated sludge and EUBmix (FLUOS; green). Bar = 10 μm.

PAOs/GAOs

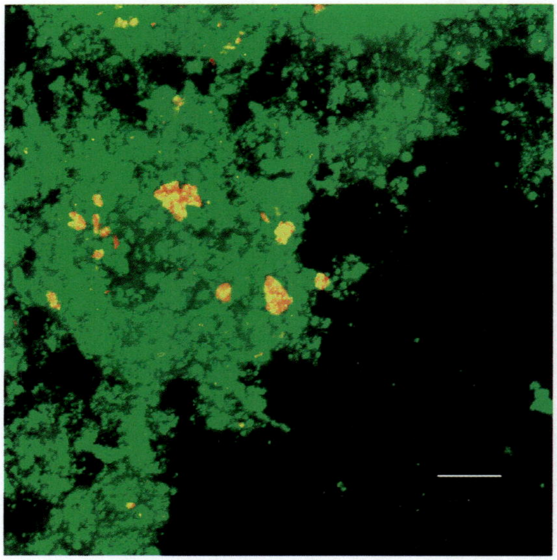

Figure 9.11. Typical cell aggregates of denitrifying *Pseudomonas* spp. detected with probe Pae997 (Cy3; red) in activated sludge and EUBmix (FLUOS; green). Bar = 10 μm.

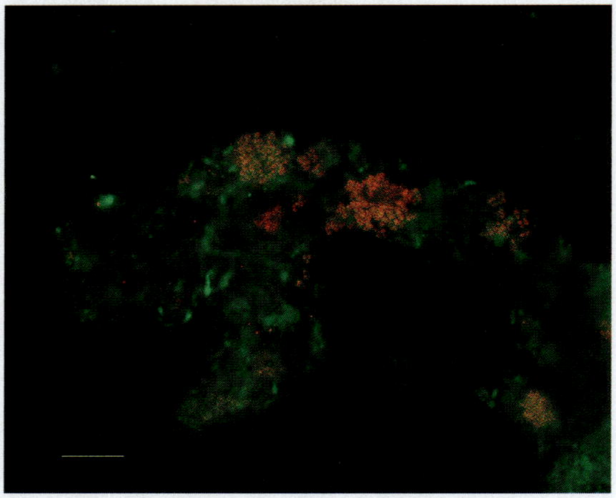

Figure 9.12. Typical cocci in tetrad aggregates of actinobacterial PAOs detected with probe Actino-221 (Cy3; red) in activated sludge and EUBmix (FLUOS; green). Bar = 10 μm.

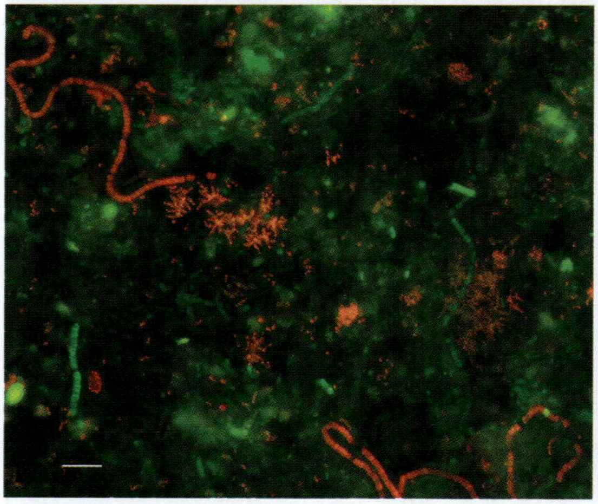

Figure 9.13. Atypical cell aggregates of actinobacterial PAOs detected with probe Actino-221 (Cy3; red) in activated sludge and EUBmix (FLUOS; green). Bar = 10 μm.

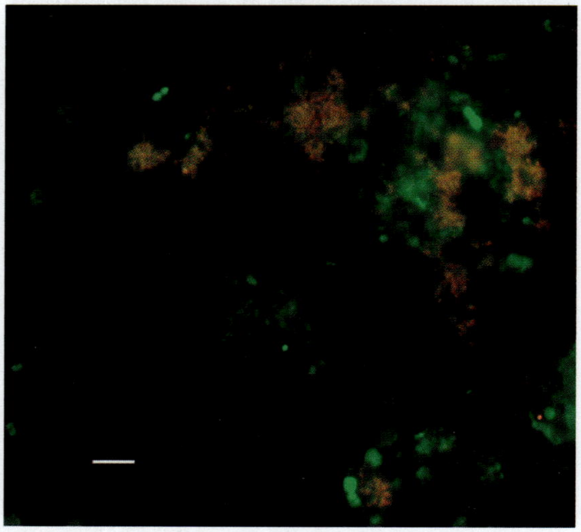

Figure 9.14. Typical cell aggregates of actinobacterial PAOs detected with probe APAO-658 (Cy3; red) in activated sludge and EUBmix (FLUOS; green). Bar = 10 μm.

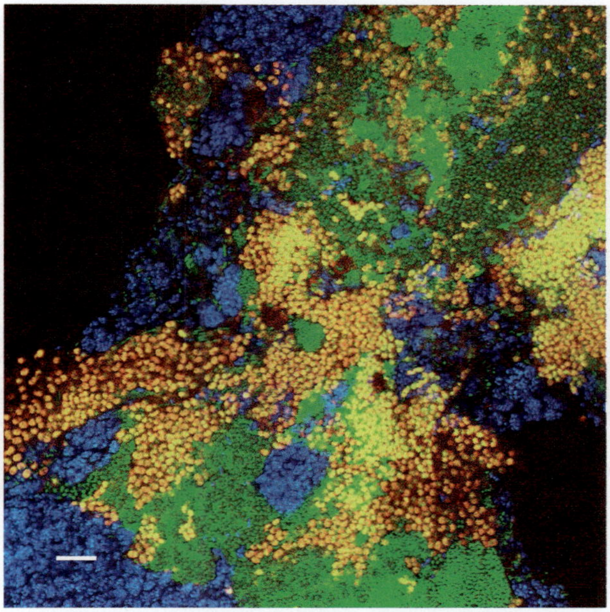

Figure 9.15. Cell aggregates of *Competibacter* GAOs and *Accumulibacter* PAOs from a lab-scale reactor detected with probe GB + GB_G1 + GB_G2 (Cy3; red) in activated sludge and PAOmix (Cy5; blue) and EUBmix (FLUOS; green). Yellow overlay of GB + GB_G1 + GB_G2 and EUBmix. Light blue overlay of PAOmix and EUBmix. Bar = 10 μm.

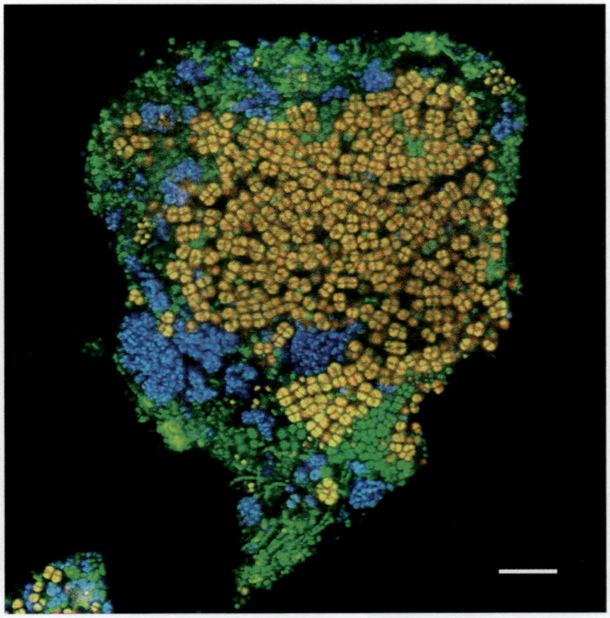

Figure 9.16. Cell aggregates of *Defluviicoccus* GAOs and *Accumulibacter* PAOs from a lab-scale reactor detected with probe DF2mix (Cy3; red) in activated sludge and PAOmix (Cy5; blue) and EUBmix (FLUOS; green). Yellow overlay of DF2mix and EUBmix. Light blue overlay of PAOmix and EUBmix. Bar = 10 μm.

Filamentous bacteria

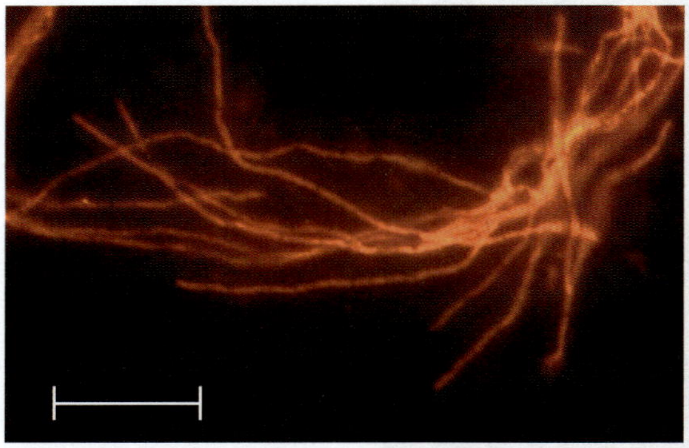

Figure 9.17. Typical *H. hydrossis* morphotype (*Bacteroidetes*) detected with probe HHY (Cy3; red) in activated sludge. Bar = 10 μm.

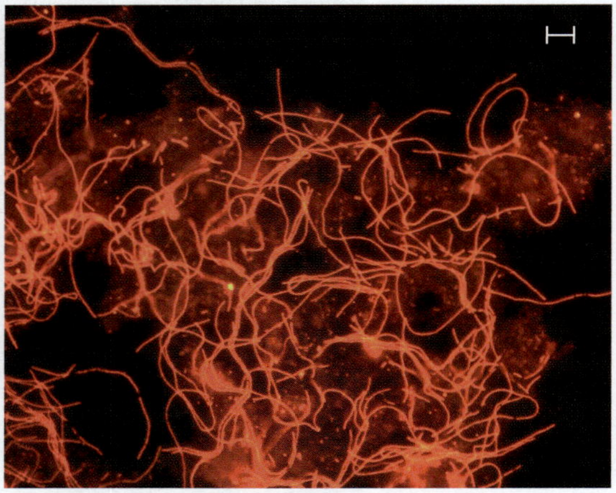

Figure 9.18. Typical *M. parvicella* morphotype (*Actinobacteria*) detected with probe MPAmix (Cy3; red) in activated sludge. Bar = 10 μm.

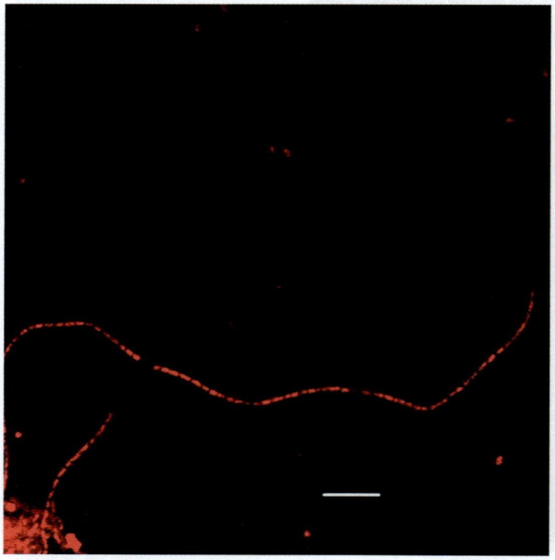

Figure 9.19. Typical *M. calida* morphotype (*Actinobacteria*) detected with probe Mpa-T1-1260 (Cy3; red) in activated sludge. Bar = 10 μm.

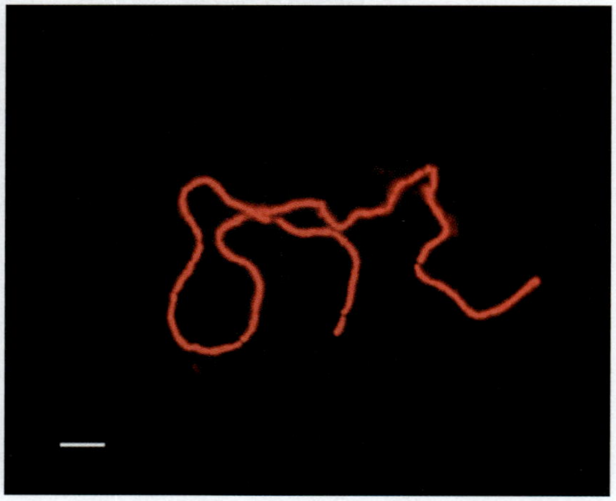

Figure 9.20. Typical *N. limicola* II morphotype (*Actinobacteria*) detected with probe NLIMII175 (Cy3; red) in activated sludge. Bar = 10 μm.

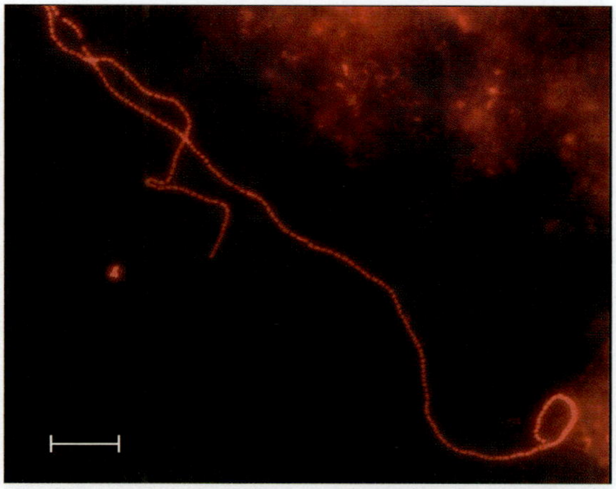

Figure 9.21. Typical *N. limicola* II morphotype (*Chloroflexi*) detected with probe AHW183 (Cy3; red) in activated sludge. Bar = 10 μm.

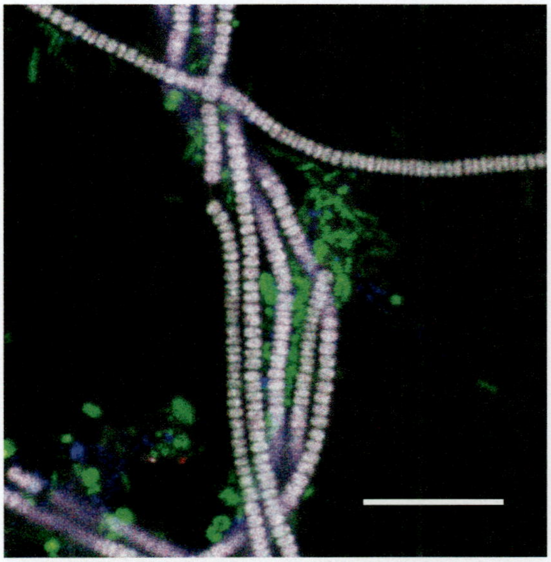

Figure 9.22. Typical *N. limicola* II morphotype *Candidatus* Monilibacter batavus (*Alphaproteobacteria*) detected with probe DF1004 (Cy3; red), DF198 (Cy5; blue) and EUBmix (FLUOS; green) in a lab-scale reactor. White overlay of DF1004, DF198 and EUBmix. Bar = 10 μm.

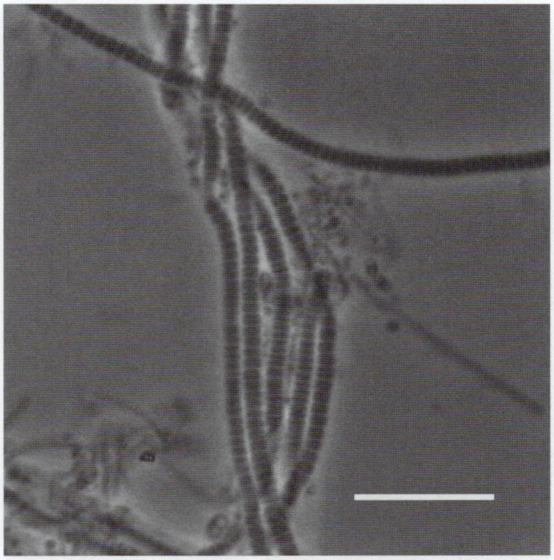

Figure 9.23. Same *Candidatus* Monilibacter batavus aside in phase contrast.

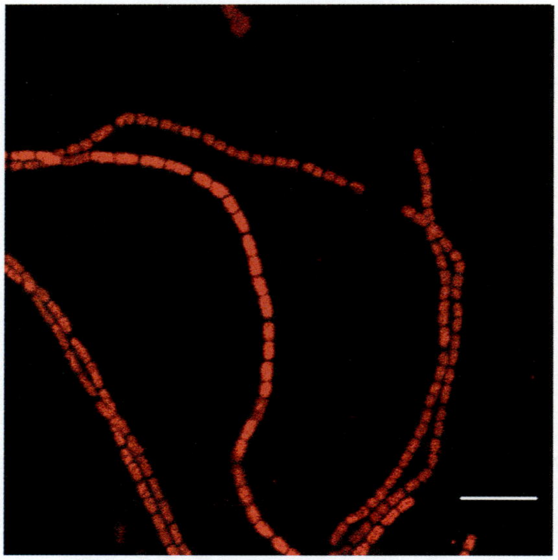

Figure 9.24. Typical *N. limicola* II morphotype *Candidatus* Alysiomicrobium bavaricum (*Alphaproteobacteria*) detected with PPx3-1428 (Cy3; red) in activated sludge. Bar = 10 μm.

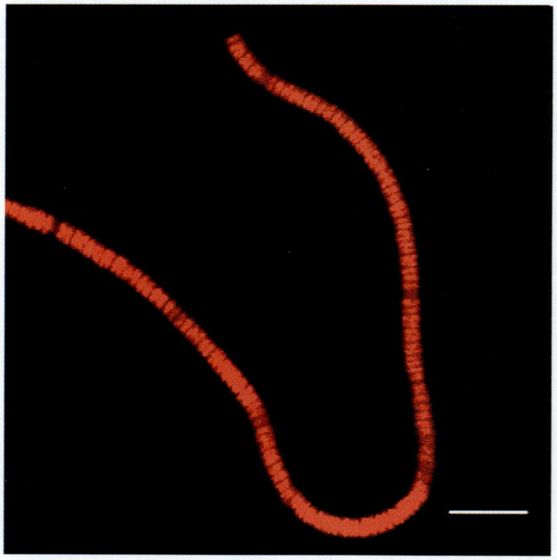

Figure 9.25. Typical *N. limicola* II morphotype *Candidatus* Alysiomicrobium bavaricum (*Alphaproteobacteria*) detected with PPx3-1428 (Cy3; red) in activated sludge. Bar = 10 μm.

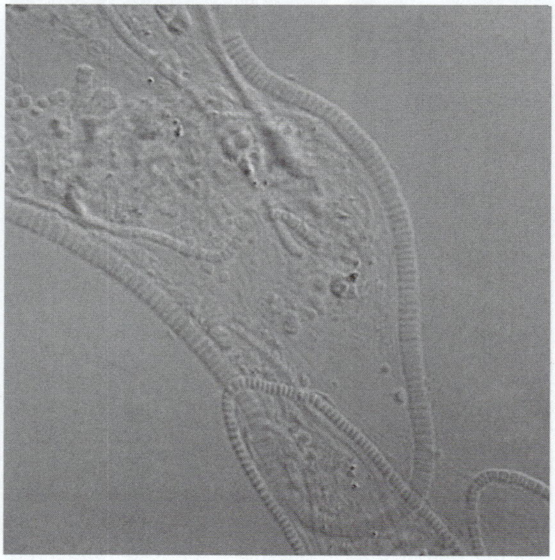

Figure 9.26. Same *Candidatus* Alysiomicrobium bavaricum aside in phase contrast.

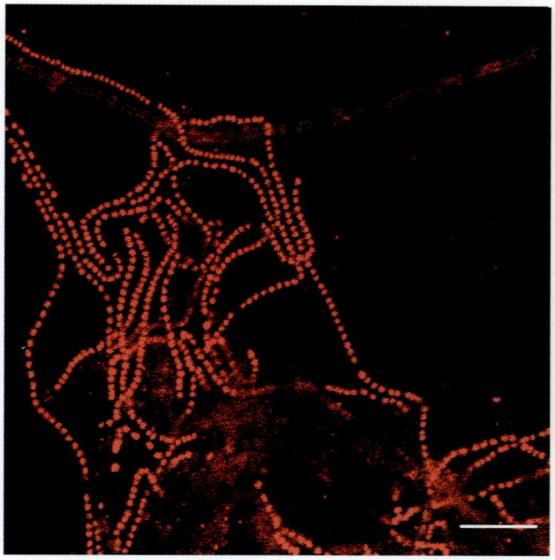

Figure 9.27. Typical *N. limicola* II morphotype *Candidatus* Sphaeronema italicum (*Alphaproteobacteria*) detected with Sita-649 + CompSita-649 (Cy3; red) in activated sludge. Bar = 10 μm.

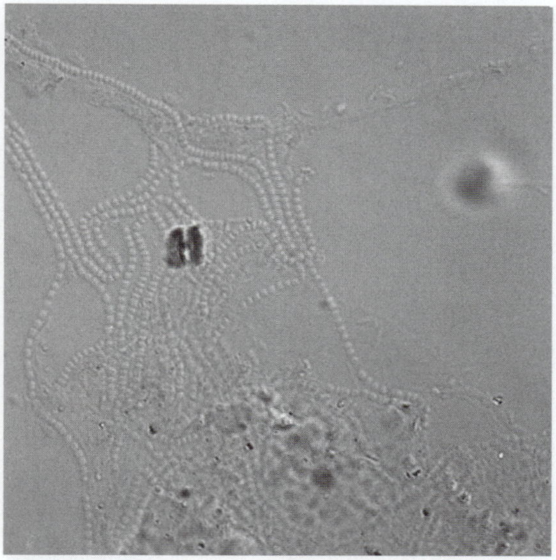

Figure 9.28. Same *Candidatus* Sphaeronema italicum aside in phase contrast.

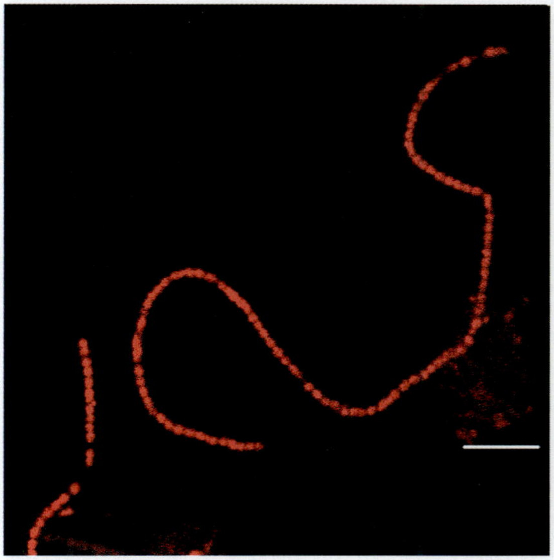

Figure 9.29. Typical *N. limicola* II morphotype *Candidatus* Alysiosphaera europaea (*Alphaproteobacteria*) detected with Noli-644 (Cy3; red) in activated sludge. Bar = 10 μm.

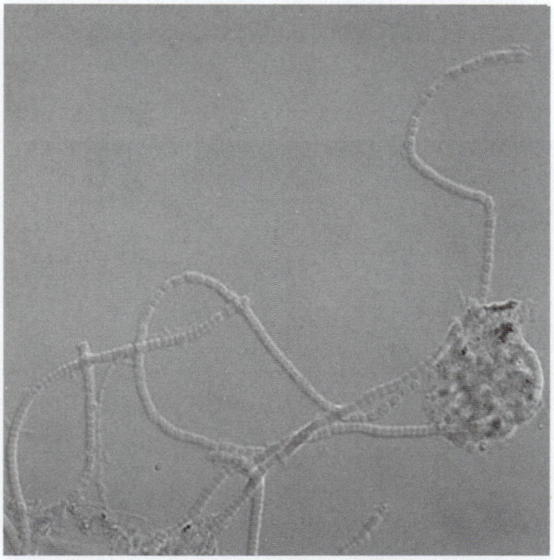

Figure 9.30. Same *Candidatus* Alysiosphaera europaea aside in phase contrast.

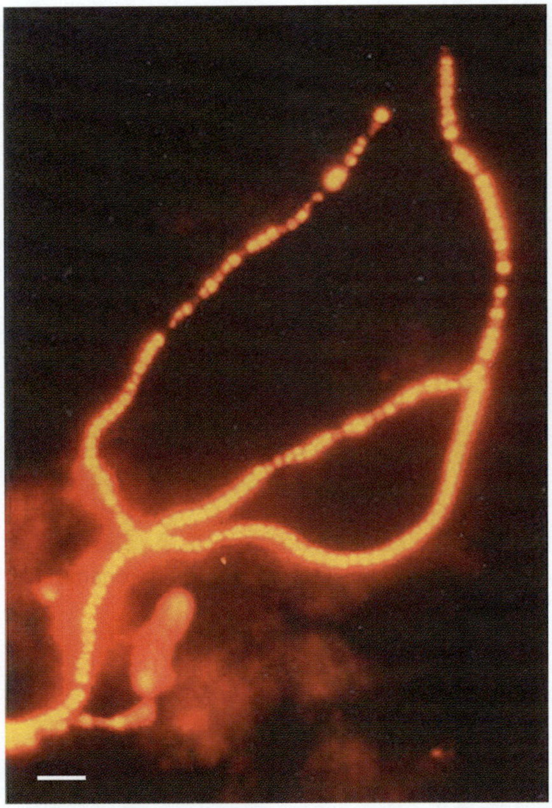

Figure 9.31. Typical *N. limicola* III morphotype *Isosphaera* spp. (*Planctomycetales*) detected with NLIMIII301 (Cy3; red) in activated sludge. Bar = 10 μm.

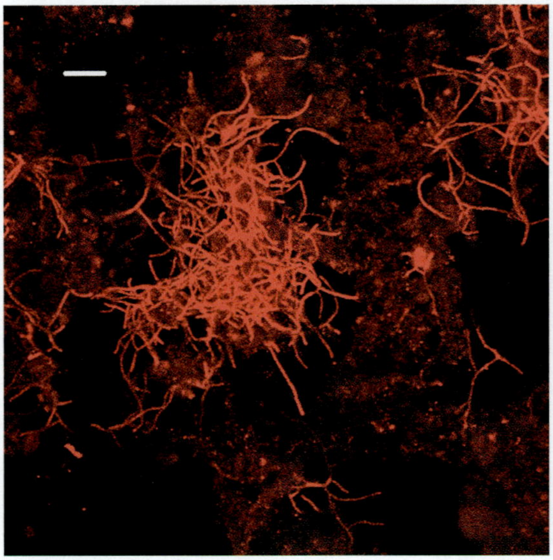

Figure 9.33. Typical Mycolata GALO morphotype *Gordonia* spp. (*Actinobacteria*) detected with Gor596 (Cy3; red) in activated sludge. Bar = 10 μm.

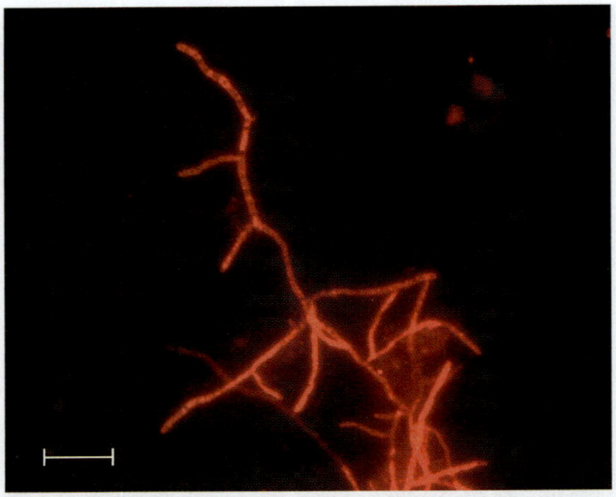

Figure 9.32. Typical Mycolata morphotype (*Actinobacteria*) detected with Myc657 (Cy3; red) in activated sludge. Bar = 10 μm.

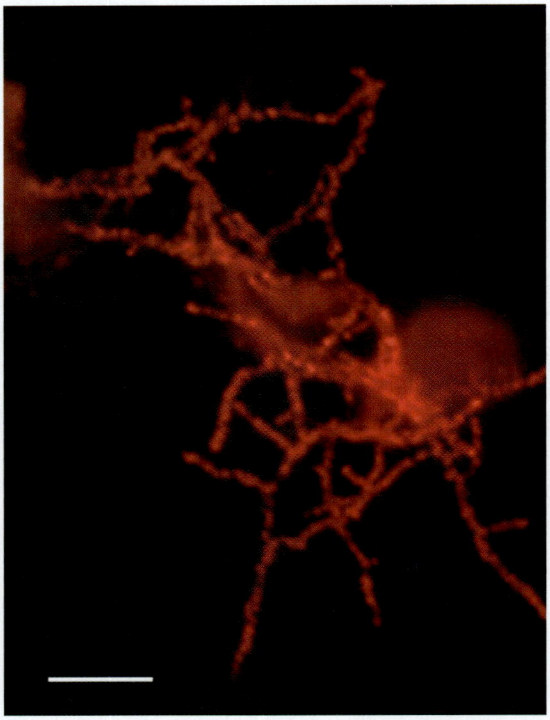

Figure 9.34. Typical Mycolata GALO morphotype *Gordonia amarae* (*Actinobacteria*) detected with G.am205 (Cy3; red) in activated sludge. Bar = 10 μm.

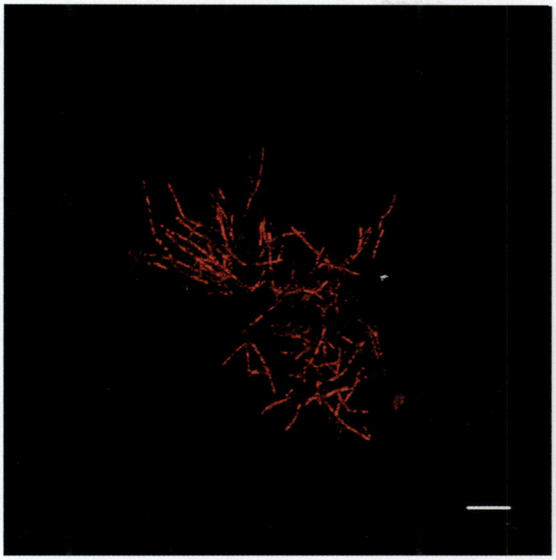

Figure 9.35. Typical Mycolata PTLO morphotype *Skermania piniformis* (*Actinobacteria*) detected with Spin1449 (Cy3; red) in activated sludge. Bar = 10 μm.

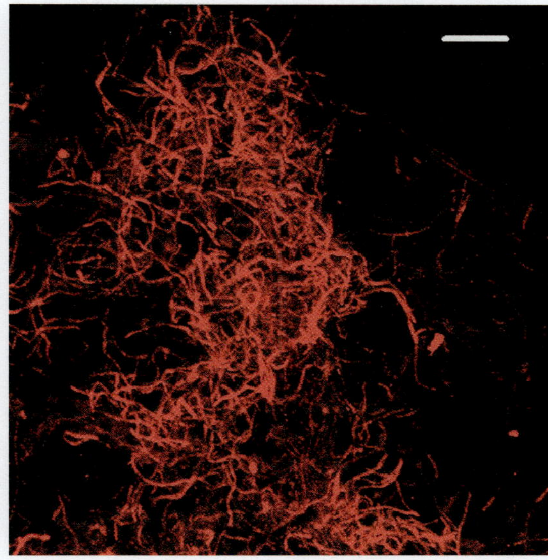

Figure 9.36. Typical Mycolata GALO morphotype *Rhodococcus erythropolis* (*Actinobacteria*) detected with R.ery619 (Cy3; red) in activated sludge. Bar = 10 μm.

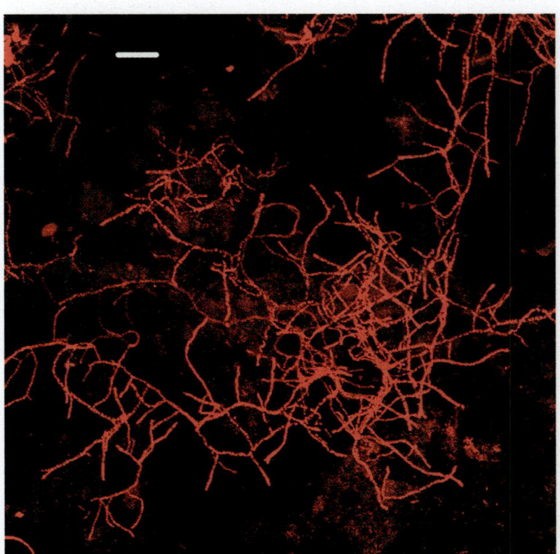

Figure 9.37. Typical Mycolata GALO morphotype *Rhodococcus* cluster B (*Actinobacteria*) detected with RHOb183 (Cy3; red) in activated sludge. Bar = 10 μm.

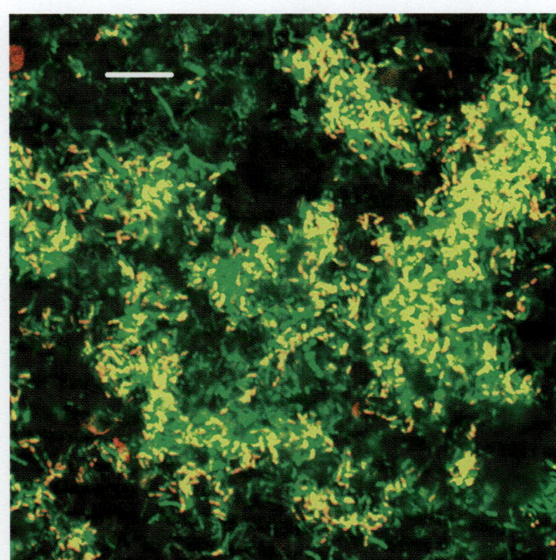

Figure 9.38. Typical Mycolata NOC morphotype *Dietzia* spp. (*Actinobacteria*) detected with DIE993 (Cy3; red) and EUBmix (FLUOS; green); Yellow overlay of DIE993 and EUBmix in activated sludge. Bar = 10 μm.

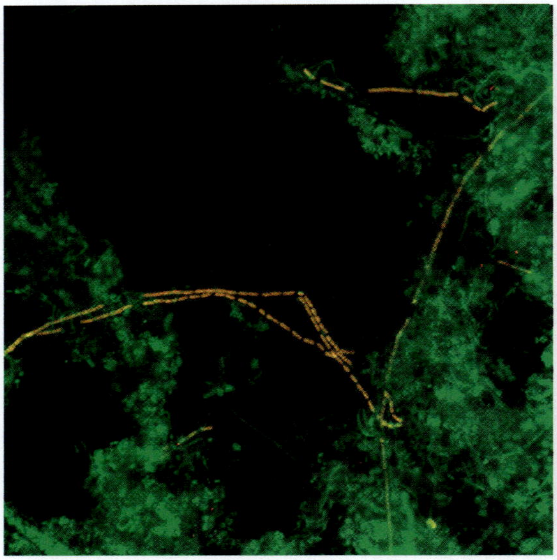

Figure 9.39. Typical *S. natans* morphotype (*Betaproteobacteria*) detected with SNA (Cy3; red) and EUBmix (FLUOS; green) in activated sludge.

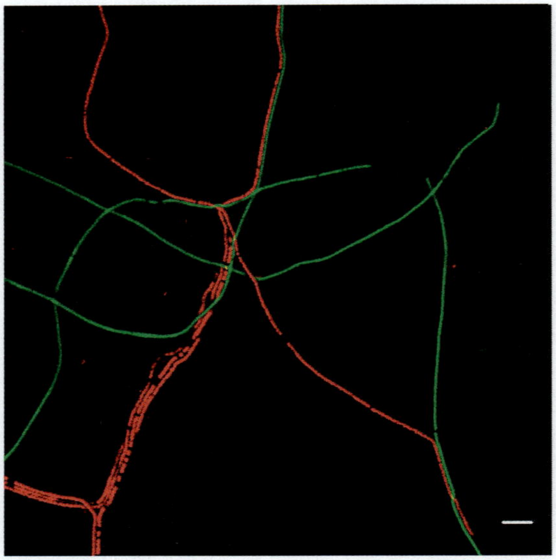

Figure 9.40. Typical *Thiothrix* morphotype (*Gammaproteobacteria*) and *Meganema perideroedes* (*Alphaproteobacteria*) detected with Meg983 + 1028 (Cy3; red) and TNI (FLUOS; green) in activated sludge. Bar = 10 μm.

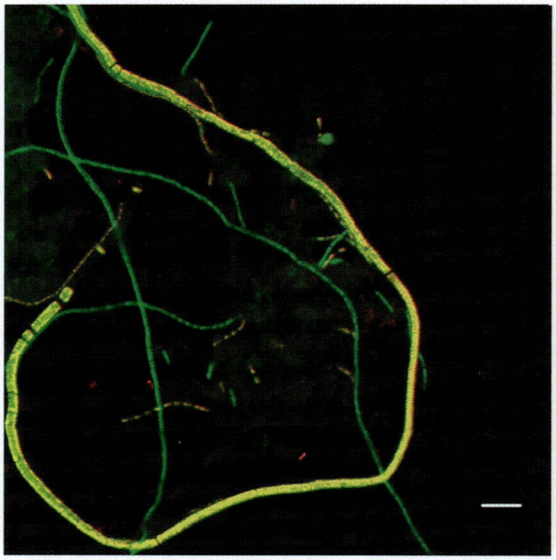

Figure 9.41. Typical *Thiothrix* morphotype (*Gammaproteobacteria*) detected with TNI (Cy3; red) and G123T (FLUOS; green) in activated sludge. Bar = 10 μm.

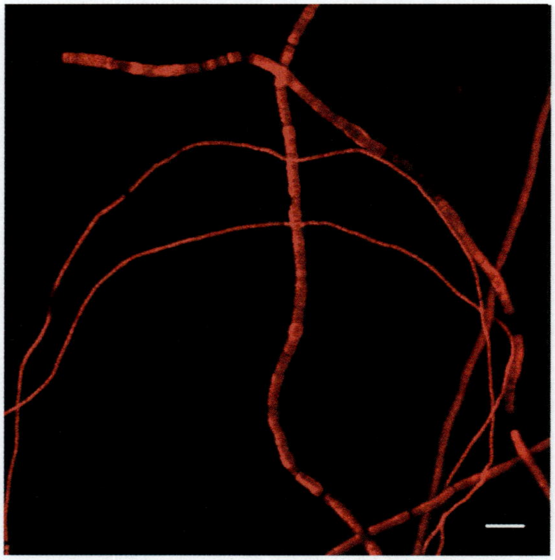

Figure 9.42. Typical 021N morphotype (*Gammaproteobacteria*) detected with G1B (Cy3; red) in activated sludge. Bar = 10 μm.

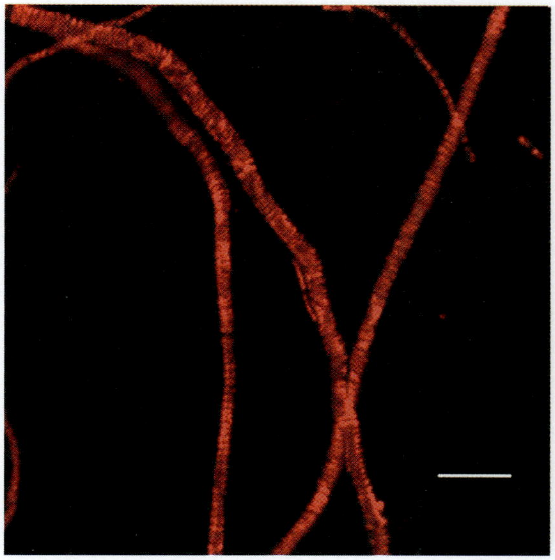

Figure 9.43. Typical 021N morphotype (*Gammaproteobacteria*) detected with 21N (Cy3; red) in activated sludge. Bar = 10 μm.

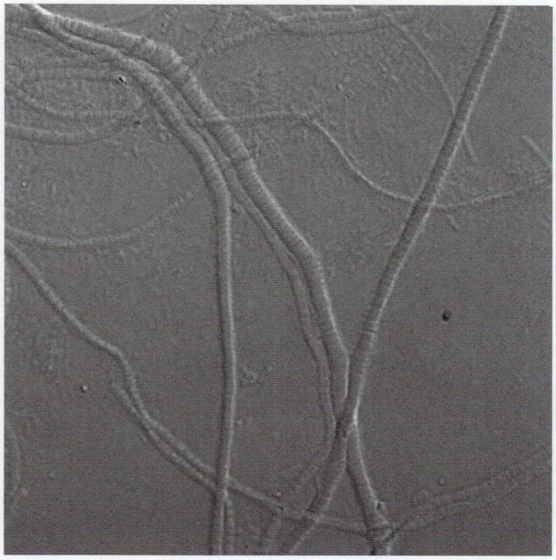

Figure 9.44. Phase contrast of the same typical 021N morphotype aside.

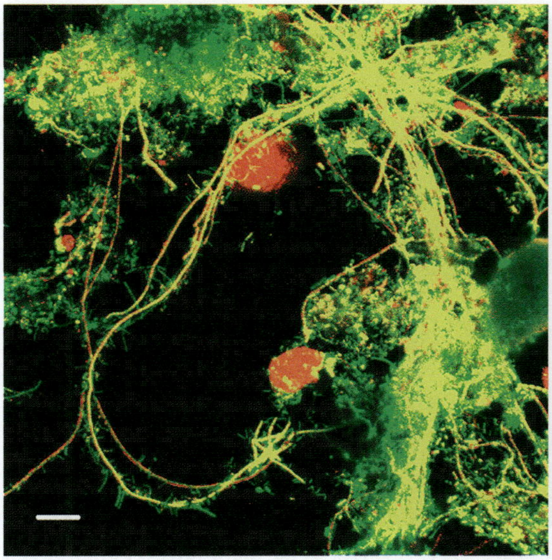

Figure 9.45. Typical 0041/0675 morphotype (*TM7*) detected with TM7305 (Cy3; red) and EUBmix (FLUOS; green) in activated sludge. Bar = 10 μm.

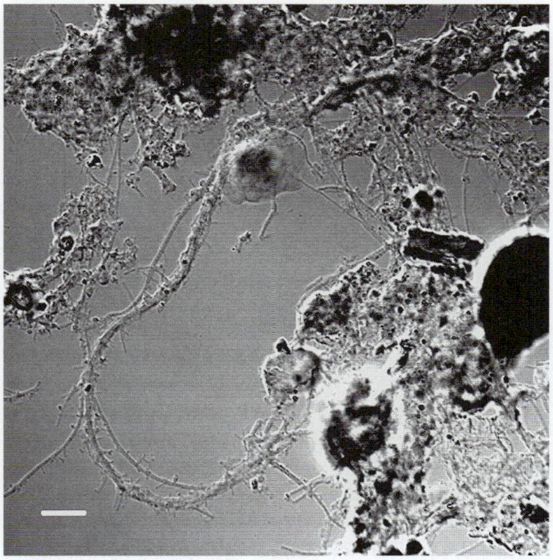

Figure 9.46. Phase contrast of the same typical 0041/0675 morphotype aside.

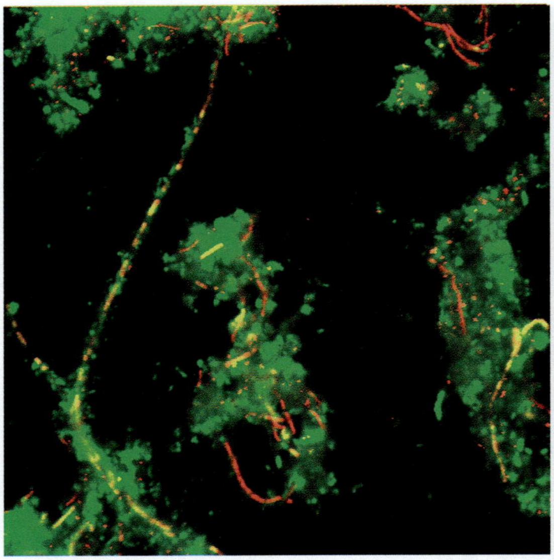

Figure 9.47. Typical 0041/0675 morphotype (*Chloroflexi*) detected with CFXmix (Cy3; red) and EUBmix (FLUOS; green), yellow overlay of CFXmix and EUBmix in activated sludge. Bar = 10 μm.

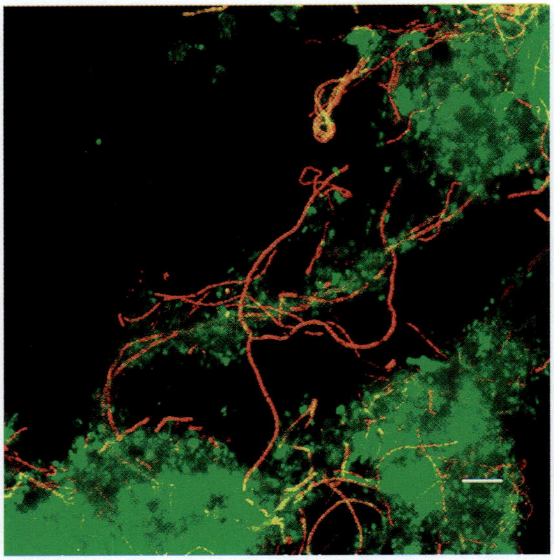

Figure 9.48. Typical 0041/0675 morphotype (*Chloroflexi*) detected with CFXmix (Cy3; red) and EUBmix (FLUOS; green) in activated sludge. Bar = 10 μm.

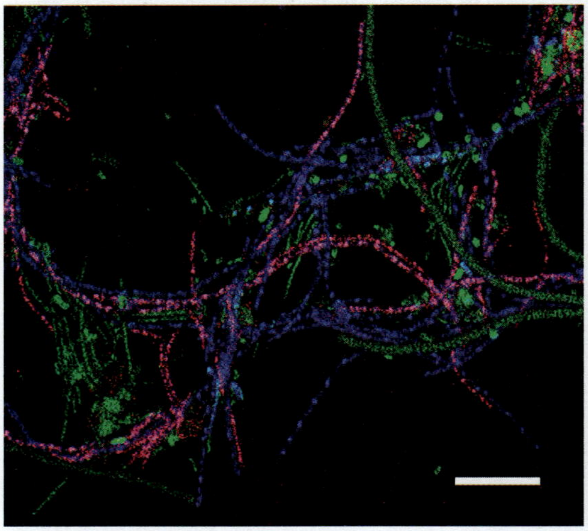

Figure 9.49. Thinner 0092 morphotype (*Chloroflexi*) detected with CFX223 (Cy3; red) and EUBmix (FLUOS; green) in activated sludge. Bar = 10 μm.

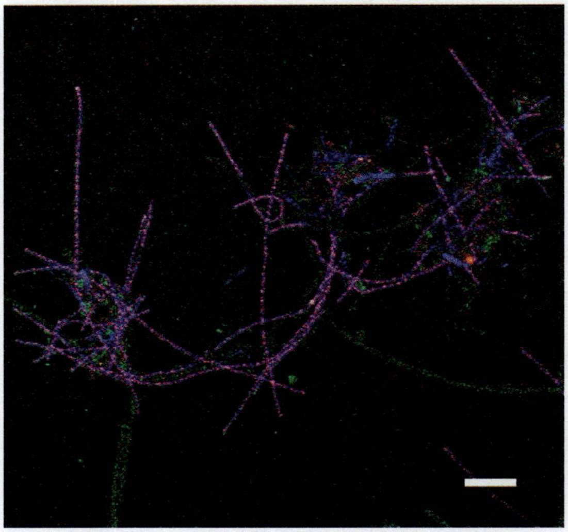

Figure 9.50. Typical 0092 morphotype (*Chloroflexi*) detected with CFX197 + competitor (Cy3; red) and EUBmix (FLUOS; green) in activated sludge. Bar = 10 μm.

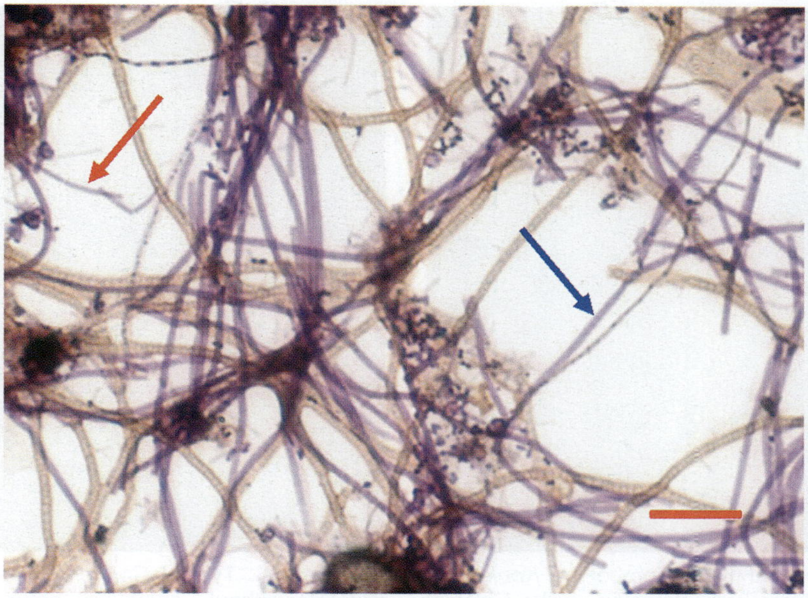

Figure 9.51. Typical 0092 morphotypes (*Chloroflexi*) stained with Neisser. The arrows indicate the two different morphotypes that can be detected with probes CFX223 and CFX197. Bar = 10 μm.

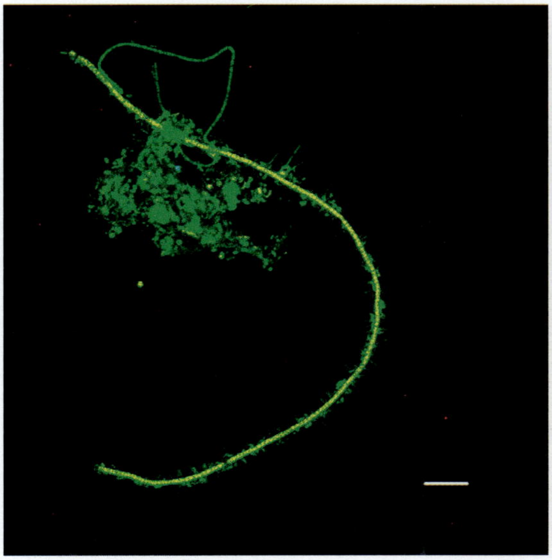

Figure 9.52. Typical 1701 morphotype *Curvibacter* spp. (*Betaproteobacteria*) detected with Curvi997 and competitors (Cy3; red) and EUBmix (FLUOS; green) in activated sludge. Bar = 10 μm.

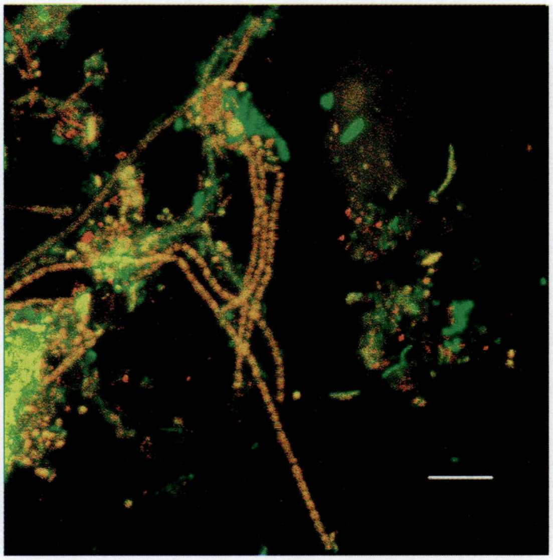

Figure 9.53. Typical 1701 morphotype *Curvibacter* spp. (*Betaproteobacteria*) detected with Curvi997 and competitors (Cy3; red) and EUBmix (FLUOS; green) in activated sludge. Bar = 10 μm.

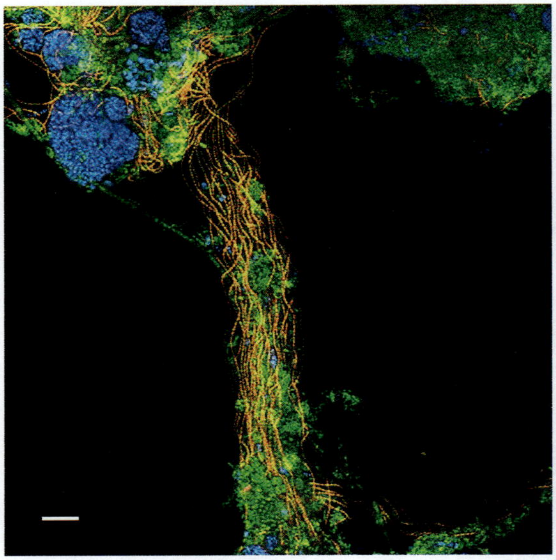

Figure 9.54. Typical 1851 morphotype *Kouleothrix aurantiaca* (*Chloroflexi*) detected with Chl1851 (Cy3; red) and EUBmix (FLUOS; green) in a reactor. Bar = 10 μm.

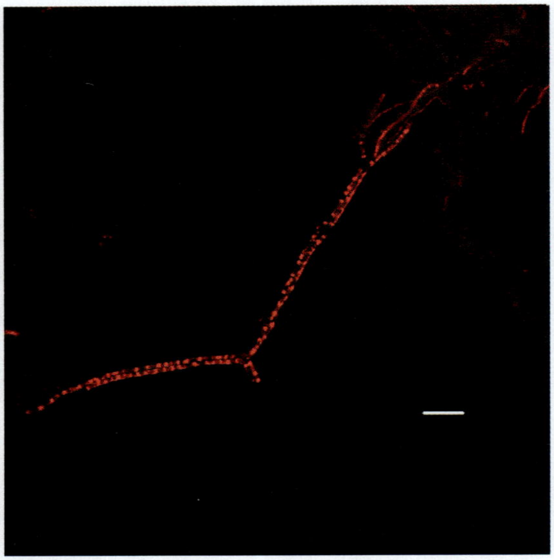

Figure 9.55. Typical *Chloroflexi* from an industrial WWTP without epiphytic growth (*Chloroflexi*) detected with CFXmix (Cy3; red) in activated sludge. Bar = 10 μm.

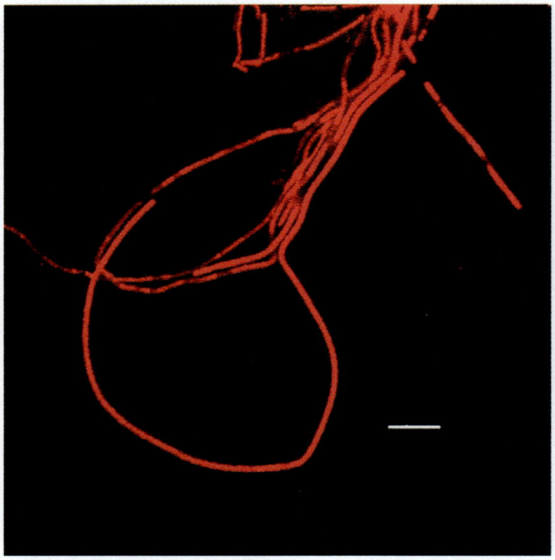

Figure 9.56. Typical *Chloroflexi* from an industrial WWTP without epiphytic growth (*Chloroflexi*) detected with CFXmix (Cy3; red) in activated sludge. Bar = 10 μm.

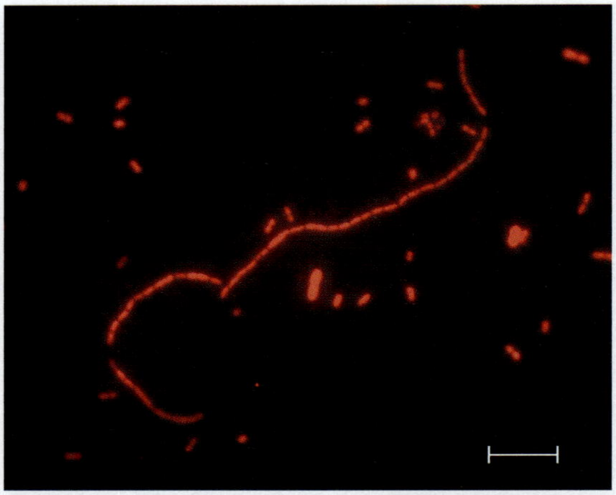

Figure 9.57. Typical 1863 morphotype *Acinetobacter* spp. (*Gammaproteobacteria*) detected with ACA23a (Cy3; red) in activated sludge. Bar = 10 μm.

10

Reference list

Adamczyk J., Hesselsoe M., Iversen N., Horn M., Lehner A., Nielsen P.H. *et al.* (2003). The isotope array, a new tool that employs substrate-mediated labeling of rRNA for determination of microbial community structure and function. *Appl. Environ. Microbiol.* 69, 6875–6887.

Alawi M., Lipski A., Sanders T., Pfeiffer E.M., and Spieck E. (2007). Cultivation of a novel cold-adapted nitrite oxidizing betaproteobacterium from the Siberian Arctic. *ISME. J.* 1, 256–264.

Amann R.I. (1995). In situ identification of micro-organisms by whole cell hybridization with rRNA-targeted nucleic acid probes. In *Molecular Microbial Ecological Manual*. Akkermans A.D.L., van Elsas J.D., and de Bruijn F.J. (eds), pp. 1–15, Kluwer Academic Publications.

Amann R.I., Binder B.J., Olson R.J., Chisholm S.W., Devereux R., and Stahl D.A. (1990). Combination of 16S rRNA-targeted oligonucleotide probes with flow cytometry for analyzing mixed microbial populations. *Appl. Environ. Microbiol.* 56, 1919–1925.

Amann R. and Fuchs B.M. (2008). Single-cell identification in microbial communities by improved fluorescence in situ hybridization techniques. *Nat. Rev. Microbiol.* 6, 339–348.

Amann R. and Ludwig W. (2000). Ribosomal RNA-targeted nucleic acid probes for studies in Microbial Ecology. *FEMS Microbiol. Rev.* 24, 555–565.

Amann R.I., Ludwig W., and Schleifer K.H. (1995). Phylogenetic identification and in situ detection of individual microbial cells without cultivation. *Microbiol. Rev.* 59, 143–169.

Amann R., Ludwig W., Schulze R., Spring S., Moore E., and Schleifer K.H. (1996). rRNA-targeted oligonucleotide probes for the identification of genuine and former pseudomonads. *Syst. Appl. Microbiol.* 19, 501–509.

Ashelford K.E., Weightman A.J., and Fry J.C. (2002). PRIMROSE: a computer program for generating and estimating the phylogenetic range of 16S rRNA oligonucleotide probes and primers in conjunction with the RDP-II database. *Nucleic Acids Res.* 30, 3481–3489.

Beer M., Seviour E.M., Kong Y., Cunningham M., Blackall L.L., and Seviour R.J (2002). Phylogeny of the filamentous bacterium Eikelboom Type 1851, and design and application of a 16S rRNA targeted oligonucleotide probe for its fluorescence in situ identification in activated sludge. *FEMS Microbiol Lett.* 207, 179–183.

Beimfohr C., Krause A., Amann R., Ludwig W., and Schleifer K.H. (1993). In-situ identification of *Lactococci, Enterococci* and *Streptococci. Syst. Appl. Microbiol.* 16, 450–456.

Bjornsson L., Hugenholtz P., Tyson G.W., and Blackall L.L. (2002). Filamentous *Chloroflexi* (green non-sulfur bacteria) are abundant in wastewater treatment processes with biological nutrient removal. *Microbiology* 148, 2309–2318.

Blackall L.L., Stratton H., Bradford D., delDot T., Sjorup C., Seviour E.M., and Seviour R.J. (1996). "*Candidatus* Microthrix parvicella", a filamentous bacterium from activated sludge sewage treatment plants. *Int. J. Syst. Bacteriol.* 46, 344–346.

Blackall L.L., Rossetti S., Christensson C., Cunningham M., Hartman P., Hugenholtz P., and Tandoi V. (1997). The characterization and description of representatives of 'G' bacteria from activated sludge plants. *Lett. Appl. Microbiol.* 25, 63–69.

Blackall L.L., Seviour E.M., Bradford D., Rossetti S., Tandoi V., and Seviour R.J. (2000). 'Candidatus *Nostocoida limicola*', a filamentous bacterium from activated sludge. *Int. J. Syst. Evol. Microbiol.* 50 Pt 2:703–9., 703–709.

Bollmann A., Bar-Gilissen M.J., and Laanbroek H.J. (2002). Growth at low ammonium concentrations and starvation response as potential factors involved in niche differentiation among ammonia-oxidizing bacteria. *Appl. Environ. Microbiol.* 68, 4751–4757.

Bottari B., Ercolini D., Gatti M., and Neviani E. (2006). Application of FISH technology for microbiological analysis: current state and prospects. *Appl. Microbiol. Biotechnol.* 73, 485–494.

Bouchez T., Patureau D., Dabert P., Juretschko S., Dore J., Delgenes P. *et al.* (2000). Ecological study of a bioaugmentation failure. *Environ. Microbiol.* 2, 179–190.

Bouvier T. and del Giorgio (2003). Factors influencing the detection of bacterial cells using fluorescence in situ hybridization (FISH): A quantitative review of published reports. *FEMS Microbiol. Ecol* 44, 3–15.

Bradford D., Hugenholtz P., Seviour E.M., Cunningham M.A., Stratton H., Seviour R.J., and Blackall L.L. (1996). 16S rRNA analysis of isolates obtained from gram-negative, filamentous bacteria micromanipulated from activated sludge. *Syst. Appl. Microbiol.* 19, 334–343.

Bradford D. (1997). *Molecular biological studies of filamentous bacteria associated with activated sludge bulking and foaming*. PhD thesis. University of Queensland, Australia.

Bradford D., Christensson C., Jakab N., and Blackall L.L. (1998). Molecular biological methods to detect "Microthrix parvicella" and to determine its abundance in activated sludge. *Water Sci. Technol.* 37, 37–45.

Burow L.C., Kong Y., Nielsen J.L., Blackall L.L., and Nielsen P.H. (2007). Abundance and ecophysiology of *Defluviicoccus* spp., glycogen-accumulating organisms in full-scale wastewater treatment processes. *Microbiology.* 153, 178–185.

Butler W.R., Floyd M.M., Brown J.M., Toney S.R., Daneshvar M.I., Cooksey R.C. *et al.* (2005). Novel mycolic acid-containing bacteria in the family *Segniliparaceae* fam. nov., including the genus *Segniliparus* gen. nov., with descriptions of *Segniliparus rotundus* sp nov and *Segniliparus rugosus* sp nov. *Int. J. Syst. Evol. Microbiol.* 55, 1615–1624.

Carr E.L., Eales K., Soddell J., and Seviour R.J. (2005). Improved permeabilization protocols for fluorescence in situ hybridization (FISH) of mycolic-acid-containing bacteria found in foams. *J. Microbiol. Methods* 61, 47–54.

Cech J.S. and Hartman P. (1990). Glucose induced break down of enhanced biological phosphate removal. *Environ. Technol.* 11, 651–656.

Cech J.S. and Hartman P. (1993). Competition between polyphosphate and polysaccharide accumulating bacteria in Enhanced biological phosphate removal systems. *Water Res.* 27, 1219–1225.

Chun J., Blackall L.L., Kang S.O., Hah Y.C., and Goodfellow M. (1997). A proposal to reclassify *Nocardia pinensis* Blackall *et al.* as *Skermania piniformis* gen. nov., comb. nov. *Int. J. Syst. Bacteriol.* 47, 127–131.

Cole J.R., Chai B., Farris R.J., Wang Q., Kulam S.A., McGarrell D.M. *et al.* (2005). The Ribosomal Database Project (RDP-II): sequences and tools for high-throughput rRNA analysis. *Nucleic Acids Res.* 33, D294-D296.

Cole J.R., Wang Q., Cardenas E., Fish J., Chai B., Farris R.J. *et al.* (2008). The Ribosomal Database Project: improved alignments and new tools for rRNA analysis. *Nucleic Acids Res.* 37, D141-D145

Crocetti G.R., Hugenholtz P., Bond P.L., Schuler A., Keller J., Jenkins D., and Blackall L.L. (2000). Identification of polyphosphate-accumulating organisms and design of 16S rRNA-directed probes for their detection and quantitation. *Appl. Environ. Microbiol.* 66, 1175–1182.

Crocetti G.R., Banfield J.F., Keller J., Bond P.L., and Blackall L.L. (2002). Glycogen-accumulating organisms in laboratory-scale and full-scale wastewater treatment processes. *Microbiology* 148, 3353–3364.

Daims H. and Wagner M. (2007). Quantification of uncultured microorganisms by fluorescence microscopy and digital image analysis. *Appl. Microbiol. Biotechnol.* 75, 237–248.

Daims H., Bruhl A., Amann R., Schleifer K.H., and Wagner M. (1999). The domain-specific probe EUB338 is insufficient for the detection of all *Bacteria*: development and evaluation of a more comprehensive probe set. *Syst. Appl. Microbiol.* 22, 434–444.

Daims H., Nielsen J.L., Nielsen P.H., Schleifer K.H., and Wagner M. (2001a). In situ characterization of *Nitrospira*-like nitrite-oxidizing bacteria active in wastewater treatment plants. *Appl. Environ. Microbiol.* 67, 5273–5284.

Daims H., Ramsing N.B., Schleifer K.H., and Wagner M. (2001b). Cultivation-independent, semiautomatic determination of absolute bacterial cell numbers in environmental samples by fluorescence in situ hybridization. *Appl. Environ. Microbiol.* 67, 5810–5818.

Daims H., Purkhold U., Bjerrum L., Arnold E., Wilderer P.A., and Wagner M. (2001c). Nitrification in sequencing biofilm batch reactors: lessons from molecular approaches. *Water Sci. Technol.* 43, 9–18.

Daims H., Stoecker K., and Wagner M. (2005). Fluorescence in situ hybridisation for the detection of prokaryotes. In *Molecular microbial ecology*. Osborn A.M. and Smith C.J. (eds), pp. 213–239, Bios-Garland, Abingdon, UK.

Daims H., Lucker S., and Wagner M. (2006). daime, a novel image analysis program for Microb. Ecol. and biofilm research. *Environ. Microbiol.* 8, 200–213.

Davenport R.J., Elliott J.N., Curtis T.P., and Upton J. (1998). In situ detection of rhodococci associated with activated sludge foams. *Antonie Van Leeuwenhoek.* 74, 41–48.

Davenport R.J., Curtis T.P., Goodfellow M., Stainsby F.M., and Bingley M. (2000). Quantitative use of fluorescent in situ hybridization to examine relationships between mycolic acid-containing actinomycetes and foaming in activated sludge plants. *Appl. Environ. Microbiol.* 66, 1158–1166.

de los Reyes F.L., Ritter W., and Raskin L. (1997). Group-specific small-subunit rRNA hybridization probes to characterize filamentous foaming in activated sludge systems. *Appl. Environ. Microbiol.* 63, 1107–1117.

de los Reyes M.F., de los Reyes F.L., Hernandez M., and Raskin L. (1998). Quantification of *Gordona amarae* strains in foaming activated sludge and anaerobic digester systems with oligonucleotide hybridization probes. *Appl. Environ. Microbiol.* 64, 2503–2512.

DeLong E.F., Wickham G.S., and Pace N.R. (1989). Phylogenetic stains: ribosomal RNA-based probes for the identification of single cells. *Science.* 243, 1360–1363.

Ding L. and Yokota A. (2004). Proposals of *Curvibacter gracilis* gen. nov., sp. nov. and *Herbaspirillum putei* sp. nov. for bacterial strains isolated from well water and reclassification of [*Pseudomonas*] *huttiensis*, [*Pseudomonas*] *lanceolata*, [*Aquaspirillum*] *delicatum* and [*Aquaspirillum*] *autotrophicum* as *Herbaspirillum huttiense* comb. nov., *Curvibacter lanceolatus* comb. nov., *Curvibacter delicatus* comb. nov. and *Herbaspirillum autotrophicum* comb. nov. *Int. J. Syst. Evol. Microbiol.* 54, 2223–2230.

Dubilier N., Giere O., Distel D.L., and Cavanaugh C.M. (1995). Characterization of chemoautotrophic bacterial symbionts in a gutless marine worm Oligochaeta, Annelida) by phylogenetic 16S rRNA sequence analysis and in situ hybridization. *Appl. Environ. Microbiol.* 61, 2346–2350.

Eales K.l., Nielsen J.L., Kragelund C., Seviour R.J., and Nielsen P.H. (2005). The *in situ* physiology of pine tree like organisms (PTLO) in activated sludge foams. *Acta hydrochim. hydrobiol.* 33, 203–209.

Eales K.l., Nielsen J.L., Seviour E.M., Nielsen P.H., and Seviour R.J. (2006). The in situ physiology of *Skermania piniformis* in foams in Australia activated sludge plants. *Environ. Microbiol.* 8, 1712–1720.

Eikelboom D.H. (1975). Filamentous organisms observed in activated sludge. *Water Res.* 9, 365–388.

Eikelboom D.H. (2000). *Process control of activated sludge plants by microscopic investigation*. IWA Publishing, London, England.

Eikelboom D.H. (2006). *Identification and control of filamentous microorganisms in industrial wastewater treatment plants*. IWA Publishing, London, England.

Eikelboom D. and Geurkink B. (2000). *Catalogue of filamentous microorganisms occurring in industrial activated sludge plants*. TNO-MEP, Report R 2001/090

Eikelboom D.H. and Geurkink B. (2002). Filamentous micro-organisms observed in industrial activated sludge plants. *Water Sci. Technol.* 46, 535–542.

Eilers H., Pernthaler J., Glockner F.O., and Amann R. (2000). Culturability and in situ abundance of pelagic bacteria from the North Sea. *Appl. Environ. Microbiol.* 66, 3044–3051.

Erhart R. (1997) In situ Analyse mikrobieller Biozönosen in Abwasserreinigungsanlagen. Erhart R. PhD Thesis, Technical University Munich, Germany.

Erhart R., Bradford D., Seviour R.J., Amann R., and Blackall L.L. (1997). Development and use of fluorescent in situ hybridization probes for the detection and identification of "Microthrix parvicella" in activated sludge. *Syst. Appl. Microbiol.* 20, 310–318.

Etchebehere C., Errazquin I., Barrandeguy E., Dabert P., Moletta R., and Muxi L. (2001). Evaluation of the denitrifying microbiota of anoxic reactors. *FEMS Microbiol. Ecol.* 35, 259–265.

Foesel B.U., Gieseke A., Schwermer C., Stief P., Koch L., Cytryn E. *et al.* (2008). *Nitrosomonas* Nm143-like ammonia oxidizers and *Nitrospira marina*-like nitrite oxidizers dominate the nitrifier community in a marine aquaculture biofilm. *FEMS Microbiol. Ecol.* 63, 192–204.

Franks A.H., Harmsen H.J., Raangs G.C., Jansen G.J., Schut F., and Welling G.W. (1998). Variations of bacterial populations in human feces measured by fluorescent in situ hybridization with group-specific 16S rRNA-targeted oligonucleotide probes. *Appl. Environ. Microbiol.* 64, 3336–3345.

Fuchs G.W. and Chen M. (1975). Microbiologicalbasis of phosphate removal in the activated sludge process for the treatment of wastewater. *Microb. Ecol.* 2, 119–138.

Fuchs B.M., Glockner F.O., Wulf J., and Amann R. (2000). Unlabeled helper oligonucleotides increase the in situ accessibility to 16S rRNA of fluorescently labeled oligonucleotide probes. *Appl. Environ. Microbiol.* 66, 3603–3607.

Galván A. and de Castro F. (2007). Relationships amon filamentous microorganisms in rotating biological contacters. *Indian J. Microbiol.* 47, 25.

Garcia-Armisen T. and Servais P. (2004). Enumeration of viable E. coli in rivers and wastewaters by fluorescent in situ hybridization. *J. Microbiol. Methods.* 58, 269–279.

Gich F., Garcia-Gil J., and Overmann J. (2001). Previously unknown and phylogenetically diverse members of the green nonsulfur bacteria are indigenous to freshwater lakes. *Arch. Microbiol.* 177, 1–10.

Gieseke A., Bjerrum L., Wagner M., and Amann R. (2003). Structure and activity of multiple nitrifying bacterial populations co-existing in a biofilm. *Environ. Microbiol.* 5, 355–369.

Gieseke A., Nielsen J.L., Amann R., Nielsen P.H., and De Beer D. (2005). In situ substrate conversion and assimilation by nitrifying bacteria in a model biofilm. *Environ. Microbiol.* 7, 1392–1404.

Ginige M.P., Hugenholtz P., Daims H., Wagner M., Keller J., and Blackall L.L. (2004). Use of stable-isotope probing, full-cycle rRNA analysis, and fluorescence in situ hybridization-microautoradiography to study a methanol-fed denitrifying microbial community. *Appl. Environ. Microbiol.* 70, 588–596.

Ginige M.P., Keller J., and Blackall L.L. (2005). Investigation of an acetate-fed denitrifying microbial community by stable isotope probing, full-cycle rRNA analysis, and fluorescent in situ hybridization-microautoradiography. *Appl. Environ. Microbiol.* 71, 8683–8691.

Giuliano L., De Domenico M., De Domenico E., Hofle M.G., and Yakimov M.M. (1999). Identification of culturable oligotrophic bacteria within naturally occurring bacterioplankton communities of the Ligurian sea by 16S rRNA sequencing and probing. *Microb. Ecol.* 37, 77–85.

Goodfellow M., Chun J., Stubbs S., and Tobili A.S. (1994). Transfer of *Nocardia amarae* Lechevalier and Lechevalier 1974 to the genus *Gordona* as *Gordona amarae* comb. nov. *Lett. Appl. Microbiol.* 19, 401–405.

Goodfellow M., Alderson G., and Chun J.S. (1998). Rhodococcal systematics: problems and developments. *Antonie Van Leeuwenhoek* 74, 3–20.

Hagman M., Nielsen J.L., Nielsen P.H., and Jansen J.L. (2008). Mixed carbon sources for nitrate reduction in activated sludge-identification of bacteria and process activity studies. *Water Res.* 42, 1539–1546.

Haleem D.A.E., von Wintzingerode F., Moter A., Moawad H., and Gobel U.B. (2000). Phylogenetic analysis of rhizosphere-associated beta-subclass proteobacterial ammonia oxidizers in a municipal wastewater treatment plant based on rhizoremediation technology. *Lett. Appl. Microbiol.* 31, 34–38.

Hess A., Zarda B., Hahn D., Haner A., Stax D., Hohener P., and Zeyer J. (1997). In situ analysis of denitrifying toluene- and m-xylene-degrading bacteria in a diesel fuel-contaminated laboratory aquifer column. *Appl. Environ. Microbiol.* 63, 2136–2141.

Hesselmann R.P., Werlen C., Hahn D., van d.M., Jr., and Zehnder A.J. (1999). Enrichment, phylogenetic analysis and detection of a bacterium that performs enhanced biological phosphate removal in activated sludge. *Syst. Appl. Microbiol.* 22, 454–465.

Heylen K., Gevers D., Vanparys B., Wittebolle L., Geets J., Boon N., and De Vos P. (2006a). The incidence of nirS and nirK and their genetic heterogeneity in cultivated denitrifiers. *Environ. Microbiol.* 8, 2012–2021.

Heylen K., Vanparys B., Wittebolle L., Verstraete W., Boon N., and De Vos P. (2006b). Cultivation of denitrifying bacteria: Optimization of isolation conditions and diversity study. *Appl. Environ. Microbiol.* 72, 2637–2643.

Heylen K., Lebbe L., and De Vos P. (2008). *Acidovorax caeni* sp nov., a denitrifying species with genetically diverse isolates from activated sludge. *Int. J. Syst. Evol. Microbiol.* 58, 73–77.

Hoshino T., Yilmaz L.S., Noguera D.R., Daims H., and Wagner M. (2008). Quantification of target molecules needed to detect microorganisms by fluorescence in situ hybridization (FISH) and catalyzed reporter deposition-FISH. *Appl. Environ. Microbiol.* 74, 5068–5077.

Howarth R., Head I.M., and Unz R.F. (1998). Phylogenetic assessment of five filamentous bacteria isolated from bulking activated sludges. *Water Sci. Technol.* 37, 303–306.

Howarth R., Unz R.F., Seviour E.M., Seviour R.J., Blackall L.L., and Pickup R.W. *et al.* (1999). Phylogenetic relationships of filamentous sulfur bacteria (*Thiothrix* spp. and Eikelboom type 021N bacteria) isolated from wastewater-treatment plants and description of *Thiothrix eikelboomii* sp. nov., *Thiothrix unzii* sp. nov., *Thiothrix fructosivorans* sp. nov. and *Thiothrix defluvii* sp. nov. *Int. J. Syst. Bacteriol.* 49, 1817–1827.

Huang W.E., Stoecker K., Griffiths R., Newbold L., Daims H., Whiteley A.S., and Wagner M. (2007). Raman-FISH: combining stable-isotope Raman spectroscopy and fluorescence in situ hybridization for the single cell analysis of identity and function. *Environ. Microbiol.* 9, 1878–1889.

Hugenholtz P. and Stackebrandt E. (2004). Reclassification of *Sphaerobacter thermophilus* from the subclass *Sphaerobacteridae* in the phylum *Actinobacteria* to the class *Thermomicrobia* (emended description) in the phylum *Chloroflexi* (emended description). *Int. J. Syst. Evol. Microbiol.* 54, 2049–2051.

Hugenholtz P., Tyson G.W., Webb R.I., Wagner A.M., and Blackall L.L. (2001). Investigation of Candidate division TM7, a recently recognized major lineage of the domain Bacteria with no known pure-culture representatives. *Appl. Environ. Microbiol.* 67, 411–419.

Jenkins D., Richard M, and Daigger G. (1993). *Manual on cases and control of activated sludge bulking and foaming.* Lewis Publishers, Chelsea, Michigan.

Jenkins D., Richard M.G., and Daigger G.T. (2004). *Manual on the causes and control of activated sludge bulking, foaming, and other solids separation problems.* IWA Publishing, CRC Press, London, UK.

Juretschko S., Timmermann G., Schmid M., Schleifer K.H., Pommerening-Roser A., Koops H.P., and Wagner M. (1998). Combined molecular and conventional analyses of nitrifying bacterium diversity in activated sludge: Nitrosococcus mobilis and Nitrospira-like bacteria as dominant populations. *Appl. Environ. Microbiol.* 64, 3042–3051.

Juretschko S., Loy A., Lehner A., and Wagner M. (2002). The microbial community composition of a nitrifying-denitrifying activated sludge from an industrial sewage treatment plant analyzed by the full-cycle rRNA approach. *Syst. Appl. Microbiol* 25, 84–99.

Kanagawa T., Kamagata Y., Aruga S., Kohno T., Horn M., and Wagner M. (2000). Phylogenetic analysis of and oligonucleotide probe development for eikelboom type 021N filamentous bacteria isolated from bulking activated sludge. *Appl. Environ. Microbiol.* 66, 5043–5052.

Kartal B., *et al.*, (2007). *Candidatus* "Anammoxoglobus propionicus" a new propionate oxidizing species of anaerobic ammonium oxidizing bacteria. *Syst. Appl. Microbiol.* 30, 39–49.

Kartal B., van Niftrik L., Rattray J., van de Vossenberg J.L., Schmid M.C., Sinninghe D.J., Jetten M.S. and Strous M. (2008). *Candidatus* 'Brocadia fulgida': an autofluorescent anaerobic ammonium oxidizing bacterium. *FEMS Microbiol. Ecol.* 63, 46–55.

Kjeldsen K.U., Joulian C., and Ingvorsen K. (2004). Oxygen tolerance of sulfate-reducing bacteria in activated sludge. *Environ. Sci. Technol.* 38, 2038–2043.

Kjeldsen K.U., Joulian C., and Ingvorsen K. (2005). Effects of oxygen exposure on respiratory activities of *Desulfovibrio desulfuricans* strain DvO1 isolated from activated sludge. *FEMS Microbiol. Ecol.* 53, 275–284.

Klatte S., Rainey F.A., and Kroppenstedt R.M. (1994). Transfer of *Rhodococcus aichiensis* Tsukamura 1982 and *Nocardia amarae* Lechevalier and Lechevalier 1974 to the Genus *Gordona* a3344w3e33s *Gordona aichiensis* Comb-Nov and *Gordona amarae* Comb-Nov. *Int. J. Syst. Bacteriol.* 44, 769–773.

Kohno T., Sei K., and Mori K. (2002). Characterization of Type 1851 organism isolated from activated sludge samples. *Water Sci. Technol.* 46, 111–114.

Kong Y.H., Beer M., Rees G.N., and Seviour R.J. (2002). Functional analysis of microbial communities in aerobic- anaerobic sequencing batch reactors fed with different phosphorus/carbon (P/C) ratios. *Microbiology-Sgm* 148, 2299–2307.

Kong Y.H., Nielsen J.L., and Nielsen P.H. (2004). Microautoradiographic study of *Rhodocyclus*-related polyphosphate accumulating bacteria in full-scale enhanced biological phosphorus removal plants. *Appl. Environ. Microbiol.* 70, 5383–5390.

Kong Y., Nielsen J.L., and Nielsen P.H. (2005). Identity and ecophysiology of uncultured actinobacterial polyphosphate-accumulating organisms in full-scale enhanced biological phosphorus removal plants. *Appl. Environ. Microbiol.* 71, 4076–4085.

Kong Y., Xia Y., Nielsen J.L., and Nielsen P.H. (2007). Structure and function of the microbial community in a full-scale enhanced biological phosphorus removal plant. *Microbiology.* 153, 4061–4073.

Kragelund C., Kong Y., van der Waarde J., Thelen K., Eikelboom D., and Tandoi V. *et al.* (2006). Ecophysiology of different filamentous *Alphaproteobacteria* species from industrial waste water treatment plants. *Microbiology* 152, 3003–3012.

Kragelund C., Levantesi C., Borger A., Thelen K., Tandoi V., Eikelboom D. *et al.* (2007a). Identity, abundance and ecophysiology of filamentous *Chloroflexi* species from activated sludge treatment plants. *FEMS Microbiol. Ecol.* 59, 671–682.

Kragelund C., Remesova Z., Nielsen J.L., Thomsen T.R., Eales K.l., Seviour R.J. *et al.* (2007b). Ecophysiology of mycolic acid containing *Actinobacteria* (Mycolata) in activated sludge foams. *FEMS Microbiol. Ecol.* 61, 174–184.

Kragelund C., Levantesi C., Borger A., Thelen K., Eikelboom D., and Tandoi V. *et al.* (2008). Identity, abundance and ecophysiology of filamentous bacteria belonging to the *Bacteroidetes* present in activated sludge plants. *Microbiology.* 154, 886–894.

Kuypers M.M., Sliekers A.O., Lavik G., Schmid M., Jorgensen B.B., Kuenen J.G. *et al.* (2003). Anaerobic ammonium oxidation by anammox bacteria in the Black Sea. *Nature.* 422, 608–611.

Lajoie C.A., Layton A.C., Gregory I.R., Sayler G.S., Taylor D.E., and Meyers A.J. (2000). Zoogleal clusters and sludge dewatering potential in an industrial activated-sludge wastewater treatment plant. *Water Environ. Res.* 72, 56–64.

Larsen P., Eriksen P.S., Lou M.A., Thomsen T.R., Kong Y.H., Nielsen J.L., and Nielsen P.H. (2006). Floc-forming properties of polyphosphate accumulating organisms in activated sludge. *Water Sci. Technol.* 54, 257–265.

Layton A.C., Karanth P.N., Lajoie C.A., Meyers A.J., Gregory I.R., Stapleton R.D. *et al.* (2000). Quantification of Hyphomicrobium populations in activated sludge from an industrial wastewater treatment system as determined by 16S rRNA analysis. *Appl. Environ. Microbiol.* 66, 1167–1174.

Lemmer H., Lind G., Muller E., Schade M., and Ziegelmayer B. (2000). Scum in activated sludge plants: Impact of non-filamentous and filamentous bacteria. *Acta hydrochim. hydrobiol.* 28, 34–40.

Levantesi C., Beimfohr C., Geurkink B., Rossetti S., Thelen K., Krooneman J. *et al.* (2004). Filamentous *Alphaproteobacteria* associated with bulking in industrial wastewater treatment plants. *Syst Appl. Microbiol* 27, 716–727.

Levantesi C., Rossetti S., Thelen K., Kragelund C., Krooneman J., Eikelboom D. *et al.* (2006). Phylogeny, physiology and distribution of '*Candidatus* Microthrix calida', a new Microthrix species isolated from industrial activated sludge wastewater treatment plants. *Environ. Microbiol.* 8, 1552–1563.

Limpiyakorn T., Kurisu F., Sakamoto Y., and Yagi O. (2007). Effects of ammonium and nitrite on communities and populations of ammonia-oxidizing bacteria in laboratory-scale continuous-flow reactors. *FEMS Microbiol. Ecol.* 60, 501–512.

Lindsay M.R., Webb R.I., Strous M., Jetten M.S., Butler M.K., Forde R.J., and Fuerst J.A. (2001). Cell compartmentalisation in planctomycetes: novel types of structural organisation for the bacterial cell. *Arch. Microbiol.* 175, 413–429.

Liu J.R. and Seviour R.J. (2001). Design and application of oligonucleotide probes for fluorescent in situ identification of the filamentous bacterial morphotype *Nostocoida limicola* in activated sludge. *Environ. Microbiol.* 3, 551–560.

Liu J.R., Burrell P., Seviour E.M., Soddell J.A., Blackall L.L., and Seviour R.J. (2000). The filamentous bacterial morphotype '*Nostocoida limicola*' I contains at least two previously described genera in the low G+C gram positive bacteria. *Syst. Appl. Microbiol.* 23, 528–534.

Liu J., Dazzo F.B., Glagoleva O., Yu B., and Jain A.K. (2001a). CMEIAS: A computer-aided system for the image analysis of bacterial morphotypes in microbial communities. *Microb. Ecol.* 41, 173–194.

Liu J.R., McKenzie C.A., Seviour E.M., Webb R.I., Blackall L.L., Saint C.P., and Seviour R.J. (2001b). Phylogeny of the filamentous bacterium '*Nostocoida limicola*' III from activated sludge. *Int. J. Syst. Evol. Microbiol.* 51, 195–202.

Liu J.R., Tanner R.S., Schumann P., Weiss N., McKenzie C.A., Janssen P.H. *et al.* (2002). Emended description of the genus *Trichococcus*, description of *Trichococcus collinsii* sp. nov., and reclassification of *Lactosphaera pasteurii* as *Trichococcus pasteurii* comb. nov. and of *Ruminococcus palustris* as *Trichococcus palustris* comb. nov. in the low-G+C gram-positive bacteria. *Int. J. Syst. Evol. Microbiol.* 52, 1113–1126.

Liu W.T., Nielsen A.T., Wu J.H., Tsai C.S., Matsuo Y., and Molin S. (2001). In situ identification of polyphosphate- and polyhydroxyalkanoate-accumulating traits for microbial populations in a biological phosphorus removal process. *Environ. Microbiol.* 3, 110–122.

Liu Y. and Liu Q.S. (2006). Causes and control of filamentous growth in aerobic granular sludge sequencing batch reactors. *Biotechnol. Adv.* 24, 115–127.

Loy A., Maixner F., Wagner M., and Horn M. (2007). probeBase–an online resource for rRNA-targeted oligonucleotide probes: new features 2007. *Nucleic Acids Res.* 35, D800-D804.

Lücker S., Steger D., Kjeldsen K.U., MacGregor B.J., Wagner M., and Loy A. (2007). Improved 16S rRNA-targeted probe set for analysis of sulfate-reducing bacteria by fluorescence in situ hybridization. *J. Microbiol. Methods.* 69, 523–528.

Ludwig W. and Schleifer K.H. (2000). How quantitative is quantitative PCR with respect to cell counts? *Syst. Appl. Microbiol.* 23, 556–562.

Lydmark P., Almstrand R., Samuelsson K., Mattsson A., Sorensson F., Lindgren P.E., and Hermansson M. (2007). Effects of environmental conditions on the nitrifying population dynamics in a pilot wastewater treatment plant. *Environ. Microbiol.* 9, 2220–2233.

Macalady J.L., Lyon E.H., Koffman B., Albertson L.K., Meyer K., Galdenzi S., and Mariani S. (2006). Dominant microbial populations in limestone-corroding stream biofilms, Frasassi cave system, Italy. *Appl. Environ. Microbiol.* 72, 5596–5609.

Maixner F., Noguera D.R., Anneser B., Stoecker K., Wegl G., Wagner M., and Daims H. (2006). Nitrite concentration influences the population structure of Nitrospira-like bacteria. *Environ. Microbiol.* 8, 1487–1495.

Manz W., Amann R., Ludwig W., Vancanneyt M., and Schleifer K.H. (1996). Application of a suite of 16S rRNA-specific oligonucleotide probes designed to investigate bacteria of the phylum *Cytophaga-Flavobacter-Bacteroides* in the natural environment. *Microbiology* 142, 1097–1106.

Manz W., Amann R., Ludwig W., Wagner M., and Schleifer K.H. (1992). Phylogenetic oligodeoxynucleotide probes for the major subclasses of *Proteobacteria* – problems and solutions. *Syst. Appl. Microbiol.* 15, 593–600.

Manz W., Wagner M., Amann R., and Schleifer K.H. (1994). In situ characterization of the microbial consortia active in 2 wastewater treatment plants. *Water Res.* 28, 1715–1723.

Manz W., Eisenbrecher M., Neu T.R., and Szewzyk U. (1998). Abundance and spatial organization of Gram-negative sulfate-reducing bacteria in activated sludge investigated by in situ probing with specific 16S rRNA targeted oligonucleotides. *FEMS Microbiol. Ecol.* 25, 43–61.

Markham N.R. and Zuker M. (2008). UNAFold: software for nucleic acid folding and hybridization. *Methods Mol. Biol.* 453:3–31, 3–31.

Martin H.G., Ivanova N., Kunin V., Warnecke F., Barry K.W., McHardy A.C. *et al.* (2006). Metagenomic analysis of two enhanced biological phosphorus removal (EBPR) sludge communities. *Nature Biotechnology* 24, 1263–1269.

Maszenan A.M., Seviour R.J., Patel B.K., Schumann P., Burghardt J., Tokiwa Y., and Stratton H.M. (2000). Three isolates of novel polyphosphate-accumulating gram-positive cocci, obtained from activated sludge, belong to a new genus, *Tetrasphaera* gen. nov., and description of two new species, *Tetrasphaera japonica* sp. nov. and *Tetrasphaera australiensis* sp. nov. *Int. J. Syst. Evol. Microbiol.* 50 Pt 2:593–603., 593–603.

Meier H., Amann R., Ludwig W., and Schleifer K.H. (1999). Specific oligonucleotide probes for in situ detection of a major group of gram-positive bacteria with low DNA G + C content. *Syst. Appl. Microbiol.* 22, 186–196.

Meyer R.L., Saunders A.M., and Blackall L.L. (2006). Putative glycogen-accumulating organisms belonging to the *Alphaproteobacteria* identified through rRNA-based stable isotope probing. *Microbiology.* 152, 419–429.

Mino T., Liu W.T., Kurisu F., and Matsuo T. (1995). Modeling glycogen storage and denitrification capability of microorganisms in enhanced biological phosphate removal processes. *Water Sci. Technol.* 31, 25–34.

Mobarry B.K., Wagner M., Urbain V., Rittmann B.E., and Stahl D.A. (1996). Phylogenetic probes for analyzing abundance and spatial organization of nitrifying bacteria. *Appl. Environ. Microbiol.* 62, 2156–2162.

Mogensen G.L., Kjeldsen K.U., and Ingvorsen K. (2005). *Desulfovibrio aerotolerans* sp. nov., an oxygen tolerant sulphate-reducing bacterium isolated from activated sludge. *Anaerobe.* 11, 339–349.

Morgan-Sagastume F., Nielsen J.L., and Nielsen P.H. (2008). Substrate-dependent denitrification of abundant probe-defined denitrifying bacteria in activated sludge. *FEMS Microbiol. Ecol.* 66, 447–461.

Moter A. and Gobel U.B. (2000). Fluorescence in situ hybridization (FISH) for direct visualization of microorganisms. *J. Microbiol. Methods* 41, 85–112.

Müller E. (2006). *Bacteria and extracellular polymeric substances in activated sludge and scum formation.* PhD thesis. TU Munich, Germany.

Müller E., Schade M., and Lemmer H. (2007). Filamentous scum bacteria in activated sludge plants: detection and identification quality by conventional activated sludge microscopy versus fluorescence in situ hybridization. *Water Environ. Res.* 79, 2274–2286.

Mussmann M., Schulz H.N., Strotmann B., Kjaer T., Nielsen L.P., Rosselló-Mora R.A. *et al.* (2003). Phylogeny and distribution of nitrate-storing *Beggiatoa* spp. in coastal marine sediments. *Environ. Microbiol.* 5, 523–533.

Nakamura K., Hiraishi A., Yoshimi Y., Kawaharasaki M., Masuda K., and Kamagata Y. (1995). *Microlunatus phosphovorus* gen-nov, sp-nov, a new Gram positive polyphosphate accumulating bacterium isolated from activated sludge. *Int. J. Syst. Bacteriol.*45, 17–22.

Nam S.W., Chun J., Kim S., Kim W., Zakrzewska-Czerwinska J., and Goodfellow M. (2003). *Tsukamurella spumae* sp. nov., a novel actinomycete associated with foaming in activated sludge plants. *Syst. Appl. Microbiol.* 26, 367–375.

Neef A., Amann R., Schlesner H., and Schleifer K.H. (1998). Monitoring a widespread bacterial group: in situ detection of planctomycetes with 16S rRNA-targeted probes. *Microbiology* 144, 3257–3266.

Neef A. (1997). *Anwendung der in situ-einzelzell-identifizierung von bakterien zur populationsanalyse in komplesen mikrobiellen biozönose*. PhD thesis. TU Munich, Germany.

Neef A., Zaglauer A., Meier H., Amann R., Lemmer H., and Schleifer K.H. (1996). Population analysis in a denitrifying sand filter: Conventional and in situ identification of *Paracoccus* spp. in methanol-fed biofilms. *Appl. Environ. Microbiol.* 62, 4329–4339.

Nielsen A.T., Liu W.T., Filipe C., Grady L., Jr., Molin S., and Stahl D.A. (1999). Identification of a novel group of bacteria in sludge from a deteriorated biological phosphorus removal reactor. *Appl. Environ. Microbiol.* 65, 1251–1258.

Nielsen J.L. and Nielsen P.H. (2002a). Enumeration of acetate-consuming bacteria by microautoradiography under oxygen and nitrate respiring conditions in activated sludge. *Water Res.* 36, 421–428.

Nielsen J.L. and Nielsen P.H. (2002b). Quantification of functional groups in activated sludge by microautoradiography. *Water Sci. Technol.* 46, 389–395.

Nielsen P.H., Kragelund C., Seviour R.J., and Nielsen J.L. (2009). Identity and ecophysiology of filamentous bacteria in activated sludge. *FEMS Microbiology Reviews* (in press).

Oehmen A., Zeng R.J., Keller J., and Yuan Z. (2007). Modeling the aerobic metabolism of polyphosphate-accumulating organisms enriched with propionate as a carbon source. *Water Environ. Res.* 79, 2477–2486.

Osaka T., Yoshie S., Tsuneda S., Hirata A., Iwami N., and Inamori Y. (2006). Identification of acetate- or methanol-assimilating bacteria under nitrate-reducing conditions by stable-isotope probing. *Microb. Ecol.* 52, 253–266.

Park H.D., Wells G.F., Bae H., Criddle C.S., and Francis C.A. (2006). Occurrence of ammonia-oxidizing archaea in wastewater treatment plant bioreactors. *Appl. Environ. Microbiol.* 72, 5643–5647.

Pernthaler A., Pernthaler J., and Amann R. (2002). Fluorescence in situ hybridization and catalyzed reporter deposition for the identification of marine bacteria. *Appl. Environ. Microbiol.* 68, 3094–3101.

Pernthaler J., Pernthaler A., and Amann R. (2003). Automated enumeration of groups of marine picoplankton after fluorescence in situ hybridization. *Appl. Environ. Microbiol.* 69, 2631–2637.

Peterson S.B., Warnecke F., Madejska J., McMahon K.D., and Hugenholtz P. (2008). Environmental distribution and population biology of *Candidatus* Accumulibacter, a primary agent of biological phosphorus removal. *Environ. Microbiol.* 10, 2692–2703.

Polz M.F., Odintsova E.V., and Cavanaugh C.M. (1996). Phylogenetic relationships of the filamentous sulfur bacterium *Thiothrix ramosa* based on 16S rRNA sequence analysis. *Int. J. Syst. Bacteriol.* 46, 94–97.

Pynaert K., Smets B.F., Wyffels S., Beheydt D., Siciliano S.D., and Verstraete W. (2003). Characterization of an autotrophic nitrogen-removing biofilm from a highly loaded lab-scale rotating biological contactor. *Appl. Environ. Microbiol.* 69, 3626–3635.

Rabus R., Wilkes H., Schramm A., Harms G., Behrends A., Amann R., and Widdel F. (1999). Anaerobic utilization of alkylbenzenes and n-alkanes from crude oil in an enrichment culture of denitrifying bacteria affiliating with the beta-subclass of Proteobacteria. *Environ. Microbiol.* 1, 145–157.

Rainey F.A., Klatte S., Kroppenstedt R.M., and Stackebrandt E. (1995). *Dietzia*, a new genus including *Dietzia maris* comb. nov., formerly *Rhodococcus maris*. *Int. J. Syst. Bacteriol.* 45, 32–36.

Rasmussen H., Bruus J.H., Keiding K., and Nielsen P.H. (1994). Observations on dewaterability and physical, chemical and microbiological changes in anaerobically stored activated sludge from a nutrient removal plant. *Water Res.* 28, 417–425.

Regnault B., Martin-Delautre S., Lejay-Collin M., Lefevre M., and Grimont P.A. (2000). Oligonucleotide probe for the visualization of *Escherichia coli/Escherichia fergusonii* cells by in situ hybridization: specificity and potential applications. *Res. Microbiol.* 151, 521–533.

Roller C., Wagner M., Amann R., Ludwig W., and Schleifer K.H. (1994). In situ probing of Gram-positive bacteria with high DNA G + C content using 23S rRNA-targeted oligonucleotides. *Microbiology* 140, 2849–2858.

Roller C., Wagner M., Amann R., Ludwig W., and Schleifer K.H. (1995). In situ probing of gram-positive bacteria with high DNA G+C content using 23S rRNA-targeted oligonucleotides. *Microbiology.* 141, 1267.

Rosselló-Mora R.A., Wagner M., Amann R., and Schleifer K.H. (1995). The abundance of *Zoogloea ramigera* in sewage treatment plants. *Appl. Environ. Microbiol.* 61, 702–707.

Rossetti S., Tomei M.C., Nielsen P.H., and Tandoi V. (2005). "*Microthrix parvicella*", a filamentous bacterium causing bulking and foaming in activated sludge systems: a review of current knowledge. *FEMS Microbiol Rev.* 29, 49–64.

Ruimy R., Boiron P., Boivin V., and Christen R. (1994). A phylogeny of the genus *Nocardia* deduced from the analysis of small-subunit ribosomal Dna-sequences, including transfer of *Nocardia-amarae* to the genus *Gordona* as *Gordona-amarae* Comb-Nov. *FEMS Microbiol. Lett* 123, 261–267.

Santos M.M., Lemos P.C., Reis M.A., and Santos H. (1999). Glucose metabolism and kinetics of phosphorus removal by the fermentative bacterium *Microlunatus phosphovorus*. *Appl. Environ. Microbiol.* 65, 3920–3928.

Schade M., Beimfohr C., and Lemmer H. (2002). Phylogenetic and physiological characterization of a "*Nostocoida limicola*"-like organism isolated from activated sludge. *Water Sci. Technol.* 46, 91–97.

Schauer M. and Hahn M.W. (2005). Diversity and phylogenetic affiliations of morphologically conspicuous large filamentous bacteria occurring in the pelagic zones of a broad spectrum of freshwater habitats. *Appl. Environ. Microbiol.* 71, 1931–1940.

Schleifer, K.H., Amann, R., Luna, G.M., Rothemund, C., Springer, N., and Dorn, S. (1992). Nucleic acid probes for the identification and in situ detection of pseudomonads. *Pseudomonas: Molecular Biology and Biotechnology*. Washington, American Society for Microbiology.

Schmid M., Twachtmann U., Klein M., Strous M., Juretschko S., Jetten M. *et al.* (2000). Molecular evidence for genus level diversity of bacteria capable of catalyzing anaerobic ammonium oxidation. *Syst. Appl. Microbiol.* 23, 93–106.

Schmid M., Schmitz-Esser S., Jetten M., and Wagner M. (2001). 16S-23S rDNA intergenic spacer and 23S rDNA of anaerobic ammonium-oxidizing bacteria: implications for phylogeny and in situ detection. *Environ. Microbiol.* 3, 450–459.

Schmid M., Walsh K., Webb R., Rijpstra W.I., Pas-Schoonen K., Verbruggen M.J. *et al.* (2003). *Candidatus* "Scalindua brodae", sp. nov., *Candidatus* "Scalindua wagneri", sp. nov., two new species of anaerobic ammonium oxidizing bacteria. *Syst. Appl. Microbiol.* 26, 529–538.

Schmid M.C., Maas, B., Dapena, A., van de Pas-Schoonen, K., van de Vossenberg, J., Kartal, B., van Niftrik, L., Schmidt, I., Cirpus, I., Kuenen, J.G., Wagner, M., Sinninghe Damste J.S., Kuypers M., Revsbech N.P., Mendez R., Jetten M.S., and Strous M. (2005). Biomarkers for in situ detection of anaerobic ammonium-oxidizing (anammox) bacteria. *Appl. Environ. Microbiol.* 71, 1677–1684.

Schramm A., De Beer D., Wagner M., and Amann R. (1998). Identification and activities in situ of *Nitrosospira* and *Nitrospira* spp. as dominant populations in a nitrifying fluidized bed reactor. *Appl. Environ. Microbiol.* 64, 3480–3485.

Schramm A., De Beer D., van den Heuvel J.C., Ottengraf S., and Amann R. (1999). Microscale distribution of populations and activities of *Nitrosospira and Nitrospira* spp. along a macroscale gradient in a nitrifying bioreactor: quantification by in situ hybridization and the use of microsensors. *Appl. Environ. Microbiol.* 65, 3690–3696.

Schuppler M., Wagner M., Schön G., and Gobel U.B. (1998). In situ identification of nocardioform actinomycetes in activated sludge using fluorescent rRNA-targeted oligonucleotide probes. *Microbiology* 144, 249–259.

Sekar R., Pernthaler A., Pernthaler J., Warnecke F., Posch T., and Amann R. (2003). An improved protocol for quantification of freshwater *Actinobacteria* by fluorescence in situ hybridization. *Appl. Environ. Microbiol.* 69, 2928–2935.

Seong C.N., Kim Y.S., Baik K.S., Lee S.D., Hah Y.C., Kim S.B., and Goodfellow M. (1999). Mycolic acid-containing actinomycetes associated with activated sludge foam. *J. Microbiol.* 37, 66–72.

Seviour E.M., Blackall L.L., Christensson C., Hugenholtz P., Cunningham M.A., Bradford D., Stratton H.M., and Seviour R.J. (1997). The filamentous morphotype Eikelboom Type 1863 is not a single genetic entity. *J. Appl. Microbiol.* 82, 411–421.

Seviour R.J. and Nielsen P.H. (2009). *Microbial Ecology of Activated Sludge*. IWA Publishing, London, UK.

Seviour R.J., Mino T., and Onuki M. (2003). The microbiology of biological phosphorus removal in activated sludge systems. *FEMS Microbiol. Rev.* 27, 99–127.

Seviour R.J., Kragelund C., Kong Y., Eales K., Nielsen J.L., and Nielsen P.H. (2008). Ecophysiology of the Actinobacteria in activated sludge systems. *Antonie Van Leeuwenhoek.* 94, 21–33.

Shannon K.E., Lee D.Y., Trevors J.T., and Beaudette L.A. (2007). Application of real-time quantitative PCR for the detection of selected bacterial pathogens during municipal wastewater treatment. *Sci. Total Environ.* 382, 121–129.

Snaidr J., Beimfohr C., Levantesi C., Rossetti S., van der W.J., Geurkink B. *et al.* (2002). Phylogenetic analysis and in situ identification of "*Nostocoida limicola*"-like filamentous bacteria in activated sludge from industrial wastewater treatment plants. *Water Sci. Technol.* 46, 99–104.

Soddell J.A., Seviour R.J., Blackall L.L., and Hugenholtz P. (1998). New foam-forming nocardioforms found in activated sludge. *Water Sci. Technol.* 37, 495–502.

Soddell J.A., Stainsby F.M., Eales K.l., Kroppenstedt R.M., Seviour R.J., and Goodfellow M. (2006). *Milisia brevis* gen. nov., sp, nov., an actinomycete isolated from activated sludge foam. *Int. J. Syst. Evol. Microbiol.* 56, 739–744.

Spieck E., Hartwig C., McCormack I., Maixner F., Wagner M., Lipski A., and Daims H. (2006). Selective enrichment and molecular characterization of a previously uncultured Nitrospira-like bacterium from activated sludge. *Environ. Microbiol.* 8, 405–415.

Stackebrandt E., Rainey F.A., and Ward Rainey N.L. (1997). Proposal for a new hierarchic classification system, *Actinobacteria* classis nov. *Int. J. Syst. Bacteriol.* 47, 479–491.

Stahl D.A. and Amann R. (1991). Development and application of nucleic acid probes. In *Nucleic acid techniques in bacterial systematics*. Stackebrandt E. and Goodfellow M. Eds. pp. 205–248, Wiley.

Stainsby F.M., Soddell J., Seviour R., Upton J., and Goodfellow M. (2002). Dispelling the "Nocardia amarae" myth: a phylogenetic and phenotypic study of mycolic acid-containing actinomycetes isolated from activated sludge foam. *Water Sci. Technol.* 46, 81–90.

Stante L., Cellamare C.M., Malaspina F., Bortone G., and Tilche A. (1997). Biological phosphorus removal by pure culture of *Lampropedia* spp. *Water Res.* 31, 1317–1324.

Strous M., Fuerst J.A., Kramer E.H., Logemann S., Muyzer G., van de Pas-Schoonen K.T. *et al.* (1999). Missing lithotroph identified as new planctomycete. *Nature.* 400, 446–449.

Strous M., Pelletier E., Mangenot S., Rattei T., Lehner A., Taylor M.W. *et al.* (2006). Deciphering the evolution and metabolism of an anammox bacterium from a community genome. *Nature.* 440, 790–794.

Tandoi V., Jenkins D., and Wanner J. (2006). *Activated sludge separation problems*. IWA Publishing, London, UK.

Teske A., Ramsing N.B., Kuver J., and Fossing H. (1996). Phylogeny of *Thioploca* and related filamentous sulfide-oxidizing bacteria. *Syst. Appl. Microbiol.* 18, 517–526.

Third K.A., Sliekers A.O., Kuenen J.G., and Jetten M.S. (2001). The CANON system (Completely Autotrophic Nitrogen-removal Over Nitrite) under ammonium limitation: interaction and competition between three groups of bacteria. *Syst. Appl. Microbiol.* 24, 588–596.

Thomsen T.R., Kjellerup B.V., Nielsen J.L., Hugenholtz P., and Nielsen P.H. (2002). In situ studies of the phylogeny and physiology of filamentous bacteria with attached growth. *Environ. Microbiol.* 4, 383–391.

Thomsen T.R., Nielsen J.L., Ramsing N.B., and Nielsen P.H. (2004). Micromanipulation and further identification of FISH-labelled microcolonies of a dominant denitrifying bacterium in activated sludge. *Environ. Microbiol* 6, 470–479.

Thomsen T.R., Blackall L.L., de Muro M.A., Nielsen J.L., and Nielsen P.H. (2006a). *Meganema perideroedes* gen. nov., sp. nov., a new filamentous alphaproteobacterium from activated sludge. *Int. J. Syst. Evol. Microbiol.* 56, 1865–1868.

Thomsen T.R., Kragelund C., and Nielsen P.H. (2006b). Identity, abundance and physiology of *Aquaspirillum*-related filamentous bacteria in activated sludge. *Water Sci. Technol.* 54, 237–245.

Thomsen T.R., Kong Y., and Nielsen P.H. (2007). Ecophysiology of abundant denitrifying bacteria in activated sludge. *FEMS Microbiol. Ecol.* 60, 370–382.

Timmermans P. and Van Haute A. (1983). Denitrification with methanol – fundamental study of the growth of denitrification capacity of hyphomicrobium Sp. *Water Res.* 17, 1249–1255.

Trebesius K., Leitritz L., Adler K., Schubert S., Autenrieth I.B., and Heesemann J. (2000). Culture independent and rapid identification of bacterial pathogens in necrotising fasciitis and streptococcal toxic shock syndrome by fluorescence in situ hybridisation. *Med. Microbiol Immunol.* 188, 169–175.

Tsai C.S. and Liu W.T. (2002). Phylogenetic and physiological diversity of tetrad-forming organisms in deteriorated biological phosphorus removal systems. *Water Sci. Technol.* 46, 179–184.

Unz R.F. and Head I.M. (2005). Genus *Thiothrix* in *Gammaproteobacteria*. In *Bergey's Manual of Systematic Bacteriology*. Garrity G.M. Eds. pp. 131–142, Springer.

van de Graaf A.A., Mulder A., de Bruijn P., Jetten M.S., Robertson L.A., and Kuenen J.G. (1995). Anaerobic oxidation of ammonium is a biologically mediated process. *Appl. Environ. Microbiol.* 61, 1246–1251.

van der Star W.R., Abma W.R., Blommers D., Mulder J.W., Tokutomi T., Strous M., Picioreanu C., and Van Loosdrecht M.C. (2007). Startup of reactors for anoxic ammonium oxidation: experiences from the first full-scale anammox reactor in Rotterdam. *Water Res.* 41, 4149–4163.

van der Waarde J., Krooneman J., Geurkink B., van der Werf A., Eikelboom D., Beimfohr C. *et al.* (2002). Molecular monitoring of bulking sludge in industrial wastewater treatment plants. *Water Sci. Technol.* 46, 551–558.

van Dongen U., Jetten M.S., and Van Loosdrecht M.C. (2001). The SHARON-Anammox process for treatment of ammonium rich wastewater. *Water Sci. Technol.* 44, 153–160.

Verstraete W., Wittebolle L., Heylen K., Vanparys B., De Vos P., and van de Wiele N.T. (2007). Microbial resource management: The road to go for environmental biotechnology. *Eng. Life Sci.* 7, 117–126.

Vester F. and Ingvorsen K. (1998). Improved most-probable-number to detect sulfate-reducing bacteria with natural media and a radiotracer. *Appl. Environ. Microbiol.* 64, 1700–1707.

Wagner A.M. and Cloete E.T. (2002). 16S rRNA sequence analysis of bacteria present in foaming activated sludge. *Syst. Appl. Microbiol.* 25, 434–439.

Wagner M. and Loy A. (2002). Bacterial community composition and function in sewage treatment systems. *Curr. Opin. Biotechnol.* 13, 218–227.

Wagner M., Amann R., Kämpfer P., Assmus B., Hartmann A., Hutzler P. *et al.* (1994a). Identification and in situ detection of Gram-negative filamentous bacteria in activated sludge. *Syst. Appl. Microbiol.* 17, 405–417.

Wagner M., Erhart R., Manz W., Amann R., Lemmer H., Wedi D., and Schleifer K.H. (1994b). Development of an rRNA-targeted oligonucleotide probe specific for the genus *Acinetobacter* and its application for in situ monitoring in activated sludge. *Appl. Environ. Microbiol.* 60, 792–800.

Wagner M., Rath G., Amann R., Koops H.P., and Schleifer K.H. (1995). In situ identification of ammonia-oxidizing bacteria. *Syst. Appl. Microbiol.* 18, 251–264.

Wagner M., Rath G., Koops H.P., Flood J., and Amann R. (1996). In situ analysis of nitrifying bacteria in sewage treatment plants. *Water Sci. Technol.* 34, 237–244.

Wagner M., Horn M., and Daims H. (2003). Fluorescence in situ hybridisation for the identification and characterisation of prokaryotes. *Curr. Opin. Microbiol.* 6, 302–309.

Wallner G., Erhart R., and Amann R. (1995). Flow cytometric analysis of activated sludge with rRNA-targeted probes. *Appl. Environ. Microbiol.* 61, 1859–1866.

Wang P., Li X.T., Xiang M.F., and Zhai Q. (2007). Characterization of efficient aerobic denitrifiers isolated from two different sequencing batch reactors by 16S-rRNA analysis. *J. Biosci. Bioeng.* 103, 563–567.

Weller R, Glöckner F.O., and Amann R. (2000). 16S rRNA-targeted oligonucleotide probes for the in situ detection of members of the phylum *Cytophaga-Flavobacterium-Bacteroides*. *Syst. Appl. Microbiol.* 23, 107–114.

Wilderer P.A., Bungartz H.J., Lemmer H., Wagner M., Keller J., and Wuertz S. (2002). Modern scientific methods and their potential in wastewater. *Water Res.* 36, 370–393.

Williams T.M. and Unz R.F. (1985). Filamentous sulfur bacteria of activated sludge: characterization of *Thiothrix, Beggiatoa,* and Eikelboom Type 021N strains. *Appl. Environ. Microbiol.* 49, 887–898.

Wong M.T., Tan F.M., Ng W.J., and Liu W.T. (2004). Identification and occurrence of tetrad-forming *Alphaproteobacteria* in anaerobic-aerobic activated sludge processes. *Microbiology-Sgm* 150, 3741–3748.

Xia Y., Kong Y., Thomsen T.R., and Halkjaer N.P. (2008). Identification and ecophysiological characterization of epiphytic protein-hydrolyzing saprospiraceae ("*Candidatus* Epiflobacter" spp.) in activated sludge. *Appl. Environ. Microbiol.* 74, 2229–2238.

Yilmaz L.S., Okten H.E., and Noguera D.R. (2006). Making all parts of the 16S rRNA of *Escherichia coli* accessible in situ to single DNA oligonucleotides. *Appl. Environ. Microbiol.* 72, 733–744.

Yoshie S., Noda N., Miyano T., Tsuneda S., Hirata A., and Inamori Y. (2001). Microbial community analysis in the denitrification process of saline-wastewater by denaturing gradient gel electrophoresis of PCR-amplified 16S rDNA and the cultivation method. *J. Biosci. Bioeng.* 92, 346–353.

Zeng R.J., Yuan Z.G., and Keller J. (2003). Enrichment of denitrifying glycogen-accumulating organisms in anaerobic/anoxic activated sludge system. *J. Biosci. Bioeng.* 81, 397–404.

Zilles J.L., Peccia J., Kim M.W., Hung C.H., and Noguera D.R. (2002). Involvement of *Rhodocyclus*-related organisms in phosphorus removal in full-scale wastewater treatment plants. *Appl. Environ. Microbiol.* 68, 2763–2769.

Zumft W.G. (1992). The denitrifying Prokaryotes. In *The Prokaryotes. A handbook on the biology of bacteria, ecophysiology, isolation, identification, applications*. Trüper H.G., Dworkin M., Harder W., and Schleifer K.H. Eds. pp. 554–582, Springer-Verlag.

Zwirglmaier K. (2005). Fluorescence in situ hybridisation (FISH) - the next generation. *FEMS Microbiol Lett.* 246, 151–158.

Index

Note: Page numbers with *f* and *t* refer to figures and tables, respectively.